ORIGINS
AND
GRAND FINALE

How the Bible and Science relate to the
Origin of Everything, Abuses of Political
Authority, and End Times Predictions

GARY HAITEL

iUniverse LLC
Bloomington

Origins and Grand Finale
How the Bible and Science relate to the Origin of Everything,
Abuses of Political Authority, and End Times Predictions

All scripture quotations are taken from the American Standard Version as first published by Thomas Nelson and sons in 1901, unless otherwise noted. All emphasis added. This translation of the American Standard Version is now in public domain.

Graphics created by author unless noted otherwise. Jesus on the cross image created by Gary Haitel and Kelti Thompson.

iUniverse books may be ordered through booksellers or by contacting:

iUniverse
1663 Liberty Drive
Bloomington, IN 47403
www.iuniverse.com
1-800-Authors (1-800-288-4677)

ISBN: 978-1-4917-3255-7 (sc)
ISBN: 978-1-4917-3256-4 (hc)
ISBN: 978-1-4917-3257-1 (e)

Library of Congress Control Number: 2014906962

Printed in the United States of America.

iUniverse rev. date: 05/14/2014

Origins and Grand Finale is a study dedicated to two of the most important people in my life—my father, Jans Haitel, and my mother, Helena Haitel-Gerdes, of Emmen, Netherlands, now both deceased. It is their commitment and dedication to Christianity and the Roman Catholic Church and their persistent search for the truth that inspired me to regenerate their pursuit of what is fact and what is myth.

Contents

Introduction

Did the universe truly originate through a big bang, or was it through some other means? Did life on Earth truly evolve from nothing, or for no reason, or is there some other means?

Common sense is another term meaning "ordinary, sensible understanding"; good, intelligent decisions and judgments cannot be made without common sense.

Logic is the art, or science, of correct reasoning, or the study of reasonable thinking.

The word "theory" normally refers to a guess or a hunch, a possibility, or a speculative idea—something that could possibly be true but is not proven. In science, it is essential to recognize that there is a difference between the terms "hypothesis," "theory," and "law of nature."

Word Net Search at Princeton University claims a hypothesis is a tentative insight into the natural world, a concept that is not yet verified but that, if true, would explain certain facts or phenomena. A theory is a well-substantiated explanation of some aspect of the natural world; an organized system of accepted knowledge that applies in a variety of circumstances to explain a specific set of phenomena. A law of nature is an explanation of some aspect of the natural world that is so firmly established that it is very unlikely to be changed in the future.

However, over the last decade, modern science has adopted a different meaning for the word "theory." A theory is now a belief that has been generally accepted by scientists as a result of actual experimentation or observation. In modern science, a theory is often considered an established and experimentally verified fact or collection of facts about the natural world. Unlike the everyday use of the word, "theory" no longer involves theoretical speculation.

Ernest Rutherford (1871–1937), a New Zealand/Canadian scientist in chemistry and physics, became known as the father of nuclear physics. In his early work he discovered the concept of radioactive half-life and proved that radioactivity involved the transmutation of one chemical element to another. In 1908, he was awarded the Nobel Prize in Chemistry "for his investigations into the disintegration of the elements, and the chemistry of radioactive substances." Mr. Rutherford said about one hundred years ago that "any theory that cannot be explained to a bartender is no damn good." What he most likely meant by this was that a theory that does not support logic or common sense is not worth anything and not worth pursuing.

In his book *A Brief History of Time*, celebrated scientist Stephan Hawking makes a serious attempt to explain some of the complexities of science and cosmology to a general audience. However, some of his theories are of such bizarre nature that the average human mind is not capable of comprehending them. Therefore, the nine million copies of *A Brief History of Time* remain likely, for the most part, unread.

The author of *Origins and grand Finale* is not a scientist or professional astronomer, although he built many telescopes in his younger years and spent countless hours observing the universe, developing in him a lifelong quest for finding the truth about origins. Gary Haitel is not pretending to be either a scientist or an astronomer. Consequently, this is a book for the taxi driver and the bartender, as well as for the intellectual.

Origins and Grand Finale covers three topics in total. In the first part of *Origins and Grand Finale*, the author is not proposing new scientific or astronomical theories; he merely examines current theories, however bizarre they may seem, and compares how they relate to the Holy Bible.

In the second part of *Origins and Grand Finale,* the author reflects on what the Bible reveals about the end of times. Although science has little to say about the end, the Bible is very explicit. The author has extensively studied the subject for two decades, and has spoken on the subject on several occasions. In *Origins and Grand Finale,* the author writes fiercely, but respectfully, to define the truth as he sees it. Corruption has existed ever since the beginning of mankind; the author has comprehensively examined political corruption toward the end of the age.

The third part of *Origins and Grand Finale*` merely focuses on what the expectations are after a person takes his or her last breath here on this earth. Does everything come to a finale—a dead end—or is there more than just being here and then just vanishing?

The Origin of Everything

There are only three ways to explain the existence of our universe: The universe has existed forever, without a beginning; the universe was created by something or someone greater than the universe itself; or the universe came into being from nothing, for no reason, without a cause. Prior to the twentieth century, very little attention was devoted to how the universe actually came into being, and most scientists of that time just assumed that the universe had existed forever—that it was eternal and static, without a reasonable explanation as to where it came from or how it got here—and no one seemed to care that much. With the lack of a reasonable understanding of how nature works, ancient explanations of the universe emerged, ranging from the Mesopotamian claim that matter represents the corpse of a slain deity, Tiamat, to the Greek principle that the physical universe existed before the gods. Ultimately, people turned to philosophy, the use of logic and common sense, to try to make sense of the world they lived in. Modern scientists make use of reason, mathematics, and experimental tests. Theologians have always believed that the universe was created by the word of God, according to Genesis, the first book of the Bible. For decades, both theologians and scientists alike have studied the Holy Scriptures and argued the concept in search for a logical and practical answer.

Back in 1959, a survey was conducted of prominent scientists across the United States, quarrying their understanding of the physical sciences. One of the questions asked was "What is the concept of the age of the universe?" More than 60 percent of the scientists questioned responded that the universe had no origin. They believed that the universe was eternal and static; they believed the steady-state universe had existed forever. Even the greatest scientist the world has ever known, Dr. Albert Einstein, first believed that the universe was static, although he later admitted this was "the biggest blunder of his life."

The origins of the universe and life on Earth have been the subject of debate for decades. Obviously, there is much at stake. From a Christian standpoint, either the God of the Bible is the Supreme Creator of the universe, or the universe spontaneously created itself from nothing for no apparent reason.

Up until about five hundred years ago, the state of the universe was generally accepted as it had been described by the Greek astronomer Ptolemy in his manuscript *Almagest* circa 150 BC. Ptolemy theorized that Earth was the stationary center of the universe; stars were embedded in a large outer sphere that rotated rapidly, approximately once a day, while the planets, the sun, and the moon were embedded in their own smaller spheres. In astronomical circles this phenomena is referred to as geocentric cosmology, which completely contradicts the theory of another Greek philosopher, Philolaus (480–385 BC), who suggested a system in which Earth, the moon, the sun, the planets, and the stars all orbit a central fire—a system referred to as heliocentric cosmology. Heliocentric cosmology was ignored for almost two thousand years, until the sixteenth century, when it was reestablished by Polish astronomer Nicholas Copernicus (1473–1543).

Up until the Copernicus discovery, Earth was considered a flat surface. Obviously, no one had ever traveled to the end of Earth or had any conception of how big Earth really was, where it ended, or what there was beyond the end. Much like the general conception of our universe today, no one knows how big it is, where it ends, or what there is beyond the edge. The vast majority of discoveries we make about our universe, the only thing we seem to discover is that it is much bigger than we previously had imagined.

After a prolonged period of study and observation, Copernicus was the first astronomer/scientist who mathematically formulated a complete heliocentric cosmology, which replaced Earth in the center of the universe with the sun. Although heliocentric cosmology had a simpler geometry than geocentric cosmology, originally it was not taken seriously, because it suggested that Earth was not the center of the universe and that it was moving.

Copernicus observed that some stars were stationary in the firmament while others seemed to move in different paths. He discovered that these moving stars were actually planets orbiting the sun. He performed his own calculations using his model of circular orbits and found that he could describe the planets' observed motions accurately. He was also able to make accurate predictions of the planets' relative distances from the sun.

Although Copernicus was convinced that his heliocentric model was an extremely important improvement to astronomy in general, he delayed publication of his results until near the end of his life, presumably because he was concerned that his work was so revolutionary that it would be rejected or lead to censure. Finally, in 1543, Copernicus's book *De revolutionibus orbium coelestium* (meaning "on the revolutions of the celestial spheres") was published shortly before he died. It is said that a copy of *De revolutionibus* was delivered to him on his deathbed shortly before he passed away, so he was able to take a brief last glimpse at his life's work. The book was widely read and accepted in Europe, and it gained enough support that it seriously threatened the geocentric model, which was virtually an article of faith in the Catholic Church. Copernicus's work was written in Latin, the language of the church and the clergy, and was thus not accessible to the common people. The church quietly tolerated his theory until 1616, when a very vocal Galileo confirmed the analysis of Copernicus's work and explained it in the language of the everyday man.

With the invention of the telescope in the late sixteenth century in the Netherlands, astronomy started to develop into a different, more modern form of science. Prior to that, the study of heavenly bodies was restricted to observations using the naked eye only, from convenient locations, such as tall buildings or higher elevations. In Venice, Italy, Galileo Galilei (1564–1642) became aware of the invention of the telescope and started building his own instruments, greatly improving upon the designs of the Dutch inventors. In 1610, Galileo was able to positively confirm Copernicus's theory after observing the moons of Jupiter and positively concluding that indeed the sun, not Earth, was at the center of the universe. Galileo also discovered the phases of the planet Venus, again proving without a shadow of a doubt that the planet Venus orbited the sun. Galileo's support of Copernicanism caused much controversy in his lifetime. The Catholic Church, as well as the majority of philosophers and astronomers, still supported the traditional geocentric view that Earth was flat and at the center of the universe. These views were based on the writings of Aristotle and Ptolemy, whose teachings became an article of faith in the Catholic Church despite the fact that their works were pagan writings.

In a letter to Johannes Kepler of August 19, 1610, Galileo complained that some of the philosophers who opposed his discoveries had refused even to look through his telescope. As a matter of fact, some considered looking through his telescope heresy. According to Karl von Gebler's 1897 book *Galileo Galilei,* Galileo wrote "I think, my Kepler, we will laugh at the extraordinary stupidity of the multitude. What do you say to the leading philosophers of the faculty here, to whom I have offered a thousand times of my own accord to show my studies, but who with the lazy obstinacy of a serpent who has eaten his fill have never consented to look at the planets, nor moon, nor telescope? Verily just as serpents close their ears, so do these men close their eyes to the light of truth? These are great matters; yet they do not occasion any surprise. People of this sort think that philosophy is a kind of a book like the Aeneid or the Odyssey, and that the truth is to be sought, not in the universe, not in nature, but (I use their own words) *by comparing text!* How you would laugh if you heard what the first philosopher of the faculty of Pisa brought against me in the presence of the Grand Duke, for he tried, now with logical arguments, now with magical adjurations, to tear down and argue the new planets out of heaven"

Two officials within the Catholic Church eventually denounced Galileo to the Roman Inquisition early in 1615, but he was cleared of any offence in February 1616. One month later, as result of the incident, the Roman Catholic Church's Congregation of the Index issued a decree suspending *De revolutionibus* until it could be "corrected," because they insisted that the Copernican doctrine was false and completely contradicting Holy Scripture. Galileo was warned to abandon his support for it, which he reluctantly promised to do.

When Copernicanism became even more controversial, Galileo traveled to Rome and attempted to convince the Church authorities not to ban his or Copernicus's ideas. He insisted that because of his observations and new discoveries, Copernicanism demanded serious consideration. In the end, however, Cardinal Bellarmine, acting on orders from the Inquisition, commended him not to "hold or defend" the idea that Earth moves and the sun is stationary at the center. However, the decree did not prevent Galileo from continuing to support heliocentric hypothesis. The same decree also prohibited him from working on any theories that suggested movement of Earth or that the sun was stationary, or to attempt to reconcile these theories with scripture. On the orders of Pope Paul V (1605–1621), a friend of Galileo, Cardinal Robert Bellarmine warned Galileo that the decree was about to be issued, and again warned him that he could not "hold or defend" the Copernican doctrine. Corrections to *De revolutionibus*, which omitted or altered nine sentences, were issued four years later, in 1620. In 1616, seventy-three years after his death, the Catholic Church finally banned Copernicus's book, and the ban was not lifted until the end of the eighteenth century.

Since the geocentric view had been the dominant scientific understanding for over one thousand years, the Catholic Church had accepted it as fact and truth. They felt that geocentric cosmology still agreed with the most advanced scientific knowledge available at that time, and the position corresponded with literal interpretations of scripture in several places, such as the following:

- "Jehovah reigneth; he is clothed with majesty; Jehovah is clothed with strength; he hath girded himself therewith: The world also is established, that it cannot be moved" (Psalm 93:1).
- "Say among the nations, Jehovah reigneth: The world also is established that it cannot be moved: He will judge the peoples with equity" (Psalm 96:10).

- "Fear before him, all the earth: the world also shall be stable, that it be not moved" (1 Chronicles 16:30).
- "*Who* laid the foundations of the earth, *that* it should not be removed forever" (Psalm 104:5).
- "The sun also ariseth, and the sun goeth down, and hasteth to his place where he arose" (Ecclesiastes 1:5).

However, regardless of the decree, Galileo continued to defend heliocentrism and insisted that the theory was not contradicting those scripture passages at all. In his defense argument, he referred to St Augustine of Hippo's position on scripture: not to take every passage literally, particularly when the scripture in question is a book of poetry and songs, not a book of science or history; and that the writers of the scriptures wrote from the perspective of the terrestrial world. From that vantage point, there has never been a day since the creation of the solar system that the sun did not rise or that the sun did not set.

He later again defended his views in his most famous work, *Dialogue Concerning the Two Chief World Systems*, published in 1632. In 1633, Galileo was again charged by the Inquisition, and this time he was ordered to stand trial on suspicion of heresy. After a bogus trial, the Inquisition found him guilty of grave suspicion of heresy for "following the position of Copernicus, which is contrary to the true sense and authority of Holy Scripture." He was charged with holding the opinion that the sun lay motionless at the center of the universe, that Earth was not at the center of the universe but orbited the sun. He was also charged with "holding and defending an opinion as probable, after it has been declared contrary to Holy Scripture." He was required to "abjure, curse and detest" those opinions, which he reluctantly did.

At the pleasure of the Inquisition, he was sentenced to formal imprisonment. The following day, his sentence was commuted to house arrest, which lasted for the rest of his life. His offending *Dialogue* was banned, and in an action not announced at the trial, publication of any of his works was forbidden, including any he might write in the future. According to popular legend, after his formal rejection Galileo allegedly muttered the rebellious phrase "And yet it moves," but there is no evidence that he actually did make this statement or anything like it.

The Galileo affair was the first major confrontation between science and theology; the fundamental truth that it is just as foolish for science to ignore the truths of the Bible, as it is for theology to disregard discoveries in science, was completely ignored. Some elements within the Catholic Church still insist today that Galileo was the troublemaker and that he caused the controversy by contradicting the Holy Scriptures. However, nothing could be further from the truth. Galileo was an astronomer and a scientist, not a theologian. He focused his attention on finding the truth about the universe and how the Supreme Creator had created the universe, not paying too much attention to how his findings would compare to the scriptures and how the Church would interpret them. His discoveries never contradicted the Holy Scriptures; perhaps the contradiction was just in the way the Church interpreted them.

The Inquisition's ban on reprinting Galileo's works was finally lifted in 1718 when permission was granted to publish an edition of his works (excluding the condemned *Dialogue*) in Florence. Then, in 1741, Pope Benedict XIV authorized the publication of an edition of Galileo's complete scientific works, which included a mildly censored version of the *Dialogue*. The Catholic Church's 1758 Index of Prohibited Books omitted the general prohibition of works defending heliocentrism but retained the specific prohibitions of the original uncensored versions of Copernicus's *De revolutionibus* and Galileo's *Dialogue Concerning the Two Chief World Systems*. All traces of official opposition to heliocentrism by the church were finally abandoned in 1835 when these works were dropped from the index. In 1939, Pope Pius XII (1939–1958), in his first speech to the Pontifical Academy of Sciences, within a few months of his election to the papacy, described Galileo as being among the "most audacious heroes of research ... not afraid of the stumbling blocks and the risks on the way, nor fearful of the funereal monuments." Finally, in 1992, international print and television news outlets published in that the Catholic Church had apparently vindicated Galileo.

In 2000, Pope John Paul II (1978–2005) issued a formal apology and begged forgiveness for all the mistakes committed by the Catholic Church over the last two thousand years, including the trial of Galileo.

Cardinal Ratzinger, who became Pope Benedict XVI on April 19, 2005, cited some current views on the Galileo affair: "… a symptomatic case that permits us to see how deep the self-doubt of the modern age, of science and technology goes today …" It is just as foolish for science to ignore the truths of the Bible, as it is for theology to disregard discoveries in science."

Galileo certainly made an impact on the astronomical community. Sadly, his discoveries were not recognized by the Catholic Church for another 350 years. He is still considered one of the greatest scientists of all times.

Between 1920 and 1970, cosmology took on a new dimension and transformed into a branch of physics. With this remarkable change, Russian astronomer Alexander Friedmann and Belgian astronomer/scientist Georges Lemaitre attempted to show that Einstein's general relativity equations offered solutions for an origin of the universe, as well as explanations for an expanding universe.

George Lemaitre was born in 1894 in Charleroi, Belgium. As a young man, he showed keen interest in both astronomy and theology. Lemaitre started his scientific career at the college of engineering in Leuven in 1913. After one year, he was drafted into the Belgian Army, where he served as an artillery officer during World War I, at which time he witnessed the first use of chemical warfare. When the war was over, he entered Maison Saint Rombout, a seminary of the Archdiocese of Malines, where he studied mathematics and science. After being ordained as a priest in 1923, he attended Cambridge University, where he continued to study math and science. One of his professors, Sir Arthur Eddington (1882–1944), was the director of the observatory. Eddington regarded Lemaitre as "a very brilliant student, wonderfully quick and clear-sighted, and of great mathematical ability." In 1924, Lemaitre went on to the United States, where he visited most of the major centers of astronomical research and received his PhD in physics from the Massachusetts Institute of Technology. In 1925, he returned to Belgium and became a part-time lecturer at the Catholic University of Leuven, where he later accepted a full-time position as professor, remaining there for the rest of his life.

During Lemaitre's study years, the astronomical community still believed in the steady-state universe, believing the universe was static and not moving, based on the studies of Isaac Newton and James C. Maxwell. When Albert Einstein first published his theory of general relativity in 1916, it was also based on a static universe that had existed forever, stable and not changing.

Lemaitre was actually the first scientist to propose the theory of an expanding universe, which is widely misattributed to American astronomer Edwin Hubble. Lemaitre was also the first to derive what is now known as Hubble's law, and he made the first estimation of the Hubble constant, which he published in 1927 in his famous paper *A homogeneous universe of constant mass and growing radius accounting for the radial velocity of extragalactic nebulae.* The paper was published in French in the Annals of the Scientific Society of Brussels, but it remained, for the most part, unnoticed.

Lemaitre based his theory of an expanding universe on observing a reddish glow, labeled as a redshift, surrounding objects outside of our galaxy, which made him believe that our universe was in motion and thus not stable and unchanging. A redshift indicates that an object is moving away from the observer, because the red hue indicates that the wavelength of the light is actually being stretched. A blue shift indicates that the object is moving toward the observer, because the blue hue indicates that the wavelength of the light is being compressed. Sound waves produce the same phenomena; a jet airplane flying toward you at low altitude and high speed produces a higher-pitched sound than when it flies away from you. When it flies toward you, it compresses the sound waves, making them shorter; when it flies away, the sound waves are stretched, making them longer, producing a lower-pitched sound. This phenomenon is called the Doppler effect. The Doppler effect is a change in frequency of sound or light waves caused by relative motion of the source and the observer.

In Lemaitre's article, which preceded Hubble's landmark article by two years, he also estimated the numerical value of what later became known as the Hubble constant. The Hubble constant is the initial value for the expansion rate. However, the data used by Lemaitre did not allow him to prove that there was an actual linear relation, which Hubble did two years later. At first Einstein was skeptical of this paper, and when Lemaitre questioned him at the 1927 Solvay Conference in Brussels, Belgium, Einstein pointed out that Russian astronomer Alexander Friedmann had also proposed a similar solution to his equations in 1922, also suggesting that the radius of the universe was increasing over time. Both Friedmann and Lemaitre designed their cosmologies based on Einstein's theories of relativity. Although Lemaitre was the first to suggest that expansion explains galactic redshift, he was not the first to publish a paper concerning the phenomenon. By February of 1922, American astronomer Vesto Slipher had already observed and measured the redshifts of forty-one galaxies. Arthur Eddington studied these redshifts and commented that "the great preponderance of positive [receding] velocities is very striking; but the lack of observations of southern nebulae is unfortunate, and forbids final conclusion." Because Einstein first refused to accept the idea of an expanding universe, Lemaitre later recalled saying to him, "Your math is correct, but your physics is abominable."

In 1930, Arthur Eddington published an extensive commentary on Lemaitre's 1927 article in the *Monthly Notices of the Royal Astronomical Society*, in which he described the paper as a "brilliant solution" to the controversial problems of cosmology. In his article, he labeled Lemaitre as a "pioneer in applying Albert Einstein's theory of general relativity to cosmology." Then, a year later, in 1931, Lemaitre published an article in the scientific journal *Nature* explaining his theory of the "primeval atom," again somewhat changing the course of cosmology. Lemaitre gained a lot of popularity, mainly because he was well acquainted with the work and discoveries of other astronomers, such as Alexander Friedmann, and he designed his theories to have testable implications that corresponded with observational data of the time. In particular, he explained the observed redshift of galaxies and the linear relation between distances and velocities. He introduced his theories at an opportune time, since Edwin Hubble would soon publish his velocity–distance relation, which strongly supported an expanding universe and received a lot of attention. Another important factor in Lemaitre's career was his association with the famous and well-respected Arthur Eddington at Cambridge University, who made sure that the Belgian got well-deserved attention in the scientific community.

In January of 1933, both Lemaitre and Einstein traveled to California together for a series of seminars. After the Belgian detailed his theory on the origin of the universe, describing it as "the Cosmic Egg exploding at the moment of creation," Einstein stood up, applauded, and said, "This is the most beautiful and satisfactory explanation of creation to which I have ever listened." However, there is some disagreement over the reporting of this quote in the newspapers of the time, and it may be very well be that Einstein was not actually referring to the "Cosmic Egg" theory as a whole, but to Lemaitre's proposal that cosmic rays may in fact be the leftover radiation of the initial explosion. Duncan Aikman, covering the occasion for the *New York Times*, spotlighted Lemaitre's view: "There is no conflict between religion and science. Lemaitre has been telling audiences over and over again in this country … His view is interesting and important not because he is a Catholic priest, not because he is one of the leading mathematical physicists of our time, but because he is both." Lemaitre's theory—the one he described as "the Cosmic Egg exploding at the moment of the creation"—was later termed a singularity. Little did he know that he had laid the foundation for one of the greatest scientific controversies for the coming century.

Lemaitre's deep Christian beliefs allowed him to easily accept the thought that the universe had a definite origin in the past, as according to the book of Genesis.

After he returned from the United States, he resumed work on his theory of the expanding universe. He published a more detailed version in the *Annals of the Scientific Society of Brussels*. This is when Lemaitre would achieve his greatest glory. Newspapers around the world called him a famous Belgian scientist and described him as the leader of the new cosmological physics.

However, after Lemaitre's college friend Arthur Eddington died in 1944, Cambridge University became the center of opposition to Lemaitre's theory of the cosmic egg. In fact, it was Fred Hoyle (1915–2010), astronomer at Cambridge, who sarcastically coined the term "big bang," referring to Lemaitre's theory, in a BBC radiobroadcast in March of 1949.

Today the big bang is the scientific community's most widely accepted theory attempting to explain the origin of the universe through a pure, natural process without divine intervention. Once it became generally accepted that the universe had a definite beginning, scientists began to develop theories as to how, what and when it developed and what existed prior to its beginning. The theory assumes that all of the matter and energy present in the entire universe started as a single point of infinite density and temperature known as a singularity. Modern science presumes that approximately 13.7 billion years ago this singularity, much smaller than the size of a marble, experienced a rapid inflation of matter, energy, space, and time that eventually evolved and self-organized into stars, galaxies, and planets. The big bang was not an explosion in the conventional sense of the term; it was more like a rapid expansion of space and time. However, just like an explosion, it must have been highly energetic and chaotic.

Although Fred Hoyle never publicly opposed Lemaitre and Hubble's theories of an expanding universe, he strongly disagreed with the way it was explained. As an atheist, he realized that if the universe had a definite beginning, then there also had to be a definite beginner and consequently a creator, which he found philosophically unacceptable. Instead, Hoyle, along with Thomas Gold and Hermann Bondi (with whom he had worked on radar in World War II), continued to pursue his theory for a steady-state universe. In the theories these three developed, they tried to explain the possibility of how the universe could be eternal and essentially unchanging while the galaxies they observed were still moving away from each other. Their steady-state theory was based on the creation of matter between galaxies over time, so that even though galaxies moved farther apart, new ones would develop between them to fill the space they left. The resulting universe would be in a steady state, as in the case of a flowing river. where the individual water molecules move away but the river itself remains at the same location. In a steady-state universe, galaxies would just move around in a confined space, thus not expanding the universe.

In addition to his work in astronomy, Hoyle was a writer of science fiction, publishing a number of books co-written with his son Geoffrey. Hoyle spent most of his working life at the Institute of Astronomy at Cambridge and served as its director for a number of years. He died in Bournemouth, England, after a series of strokes in 2001.

Lemaitre's supposed evidence and the discovery of the cosmic microwave background radiation in the 1960s resulted in the big bang's victory over the steady-state model—at least in the minds of most cosmologists.

Edwin Hubble (1889–1953) was an American astronomer who, independently of Lemaitre, also changed our understanding of the universe by discovering other galaxies besides our own, the Milky Way. Just like Lemaitre, he also discovered that the degree of redshift observed in the light spectra from other galaxies increased in proportion to each particular galaxy's distance from Earth. This relationship became known as Hubble's law, and it seemed indeed to confirm the expansion of the universe.

Edwin Hubble's arrival at Mount Wilson, California, in 1919 coincided roughly with the completion of the one-hundred-inch (2.5 m) Hooker Telescope, then the world's largest telescope, when it was believed that the entire universe consisted only of the Milky Way Galaxy. Systematically making observations through this newly constructed telescope, Hubble discovered distant objects and realized that these were not stars belonging to our Milky Way. They were actually other distant galaxies, first called nebulae, and like our own Milky Way, they also consisted of billions of stars. Observations he made during 1922 and 1923 proved conclusively that these nebulae were much too distant to be part of the Milky Way, meaning that the universe was actually much bigger and more massive than previously thought. His contemporary astronomers initially did not accept his idea; however, when he announced his discovery on January 1, 1925, it quickly changed the fundamental understanding of the universe. Hubble developed the most commonly used system for classifying galaxies, grouping them according to their appearance in photographic images. He arranged the different groups of galaxies in what became known as the Hubble sequence. Astronomers use his system even today. First Hubble divided the galaxies into two general categories: elliptical and spiral galaxies. Elliptical galaxies are shaped like ellipses, and spiral galaxies are shaped like spirals, with arms winding in to a bright center.

Light, when viewed through a prism spectroscope (a prism is a wedge-shaped piece of glass used to split light into a spectrum) produces spectral lines, which appear in two different types: emission lines, which are light on dark, and absorption lines, which are dark on light. When the frequencies do not line up correctly, the spectrum is shifted, with red toward the longer end and blue toward the shorter end. This change in frequency was originally considered to be caused by motion. The redshift of a distant galaxy or quasar is easily measured by comparing its spectrum with a laboratory reference spectrum. By measuring the locations of these lines in astronomical spectra, astronomers can determine the amount of redshift of the receding sources.

Using a prism spectroscope, Hubble confirmed these galactic redshifts, as observed by Lemaitre two years earlier. He suggested that these galaxies were actually moving away from us with a velocity proportional to their distance from us; the further galaxies were away, the faster they seemed to be moving. Then he correlated the distances of the galaxies with their velocities. Although Hubble continued to have reservations about the theory, the scientific community quickly accepted the theory as fact, and it became known as Hubble's law. No cause of galactic redshift other than velocity away from an observer was considered possible, so Hubble's results were accepted to mean just that—the farther a galaxy is away from us, the faster it moves away from us. Consequently, the universe had to be expanding at an incredible rate. Hubble's law became the primary explanation of galactic redshift, confirming expanding space-time, on which the big bang theory is based.

Interestingly, in 1917 Albert Einstein produced a model of space based on an expanding universe. He claimed that space was curved and distorted by gravity, and therefore it must be able to expand or contract; but after considerable deliberation, he found this assumption to farfetched. He modified his way of thinking and continued his concept of the steady-state universe, a universe that was static and not moving. After the publication of Hubble's discoveries, he promptly admitted that second-guessing his original findings was the biggest blunder of his life and quickly and publicly endorsed Lemaitre and Hubble's theory, causing both the theory and its originators to get fast recognition, and he even visited Hubble personally to thank him in 1931.

Hubble published two classic books, *The Realm of the Nebulae* in 1936 and *The Observational Approach to Cosmology* in 1937. Edwin Hubble was apparently not a religious man. Neither his books nor his biographers discuss any reflection of the Bible or biblical Creation.

On March 6, 2008, the United States Postal Service released a forty-one-cent stamp honoring Edwin Hubble. On a sheet titled American Scientists, his citation reads, "Often called a 'pioneer of distant stars', astronomer Edwin Hubble played a pivotal role in deciphering the vast and complex nature of the universe. His meticulous study of spiral nebulae proved the existence of galaxies other than our own Milky Way. Had he not died suddenly in 1953, Hubble would have won that year's Nobel Prize in Physics."

From Hubble's model, in 1948, Russian astronomer George Gamow predicted that if there actually had been a big bang, then there also must be an existence of remnants, or cosmic microwave background (CMB) radiation. The CMB was actually discovered in the 1960s and made the big bang theory more credible than its chief rival, the steady-state, or static, universe.

In 1964, radio astronomers Arno Penzias and Robert Wilson discovered a strange microwave signal buried in their data. They attempted to filter out the signal, assuming that it was merely unwanted noise, which they initially believed to be caused by pigeons. However, they soon realized that they had actually discovered a 270.425 degree

Celsius cosmic microwave background (CMB) radiation, which was present throughout the entire observable universe. George Gamow first developed this theory in 1948, but very few astronomers or scientists took it very seriously. The theory states that, after the initial big bang, the universe cooled, matter gradually stopped moving, and the gravitational forces of mass, energy, and density came to dominate the force of radiation. (Gravity is the gravitational force at the surface of a planet or other body that pulls mass toward its center) After about three hundred thousand years, the electrons and nuclei combined into atoms (mostly hydrogen), causing the radiation to decouple from matter and continue through space largely uninterrupted. This radiation is referred to as CMB, the cosmic microwave background.

The discoveries of galactic redshift and cosmic microwave background are regarded as the primary confirmation that the universe did have a definite beginning. Arno Allen Penzias (born in Munich, West Germany, 1933) and Robert Woodrow Wilson (born 1936) were both awarded a quarter of the Nobel Prize in Physics for their discoveries in 1978.

By the early 1970s, more than forty years ago, scientific consensus was reached that the big bang was indeed the way it all must have happened. Once it became generally accepted that the universe had a definite beginning, scientists began to develop theories of how, precisely, things must have happened. New theories were born to support the details. For example, the concept of universal expansion quickly moved the explosion away from the concept of a singularity.

However, serious trouble developed in the 1980s when it was discovered that galaxies did not have nearly enough visible mass to rotate as they did. In the 1990s, it was discovered that there was not even enough visible matter to support the experimentally measured expansion rate. More than 90 percent of the required mass was missing—just not there! The concepts of dark matter and dark energy quickly emerged to account for the discrepancy. It was also thought that dark matter was necessary to help make galaxies form, even if the seeds were already there. Cosmologists became desperate to find reasons why the universe seemed to be expanding. The most popular explanation is that some sort of force is pushing the accelerating the universe's expansion. That force is generally attributed to this mysterious dark energy. It is just like blowing up a balloon; the more the balloon is blown up, the bigger it gets, but there is an external force required to blow up the balloon. Dark energy is at the heart of one of the greatest mysteries of modern physics, but it may be nothing more than an illusion, according physicists at Oxford University. First, scientists must determine whether the universe is really expanding. How could they possibly know, if they don't even know how big the universe is, where it ends, and what "space" it is expanding into? Even with today's modern technology, we still do not know where the universe ends; as a matter of fact, we don't even know if there is an end to the universe! What could there possibly be beyond the end of the universe—nothing?

There are no good reasons to prolong belief in universal expansion, because observations at every level indicate otherwise. In 2008, Dr. Glenn Starkman made an interesting statement: "There are things we know, things we know we don't know, and then there are things we don't know we don't know."

Today there is increasing opposition to the big bang, and there are many astronomers and scientists contradicting the theory and attempting to prove it to be incorrect.

Logically, the big bang standard model fails every reasonable test of science. It fails in the face of the laws of thermodynamics and the conservation of energy, a system of belief so fundamental to physics that its defiance is simply inconceivable to serious scientists. The first law of thermodynamics asserts that matter or its energy equivalent (Einstein's $E=mc^2$) can neither be created nor destroyed under any circumstance. One of the logical outcomes of this law is that there is no new matter or energy appearing anywhere in the universe; nor is there any matter or energy being annihilated. All matter and energy in the universe is conserved. Consequently, this law is often referred to as the law of conservation of mass and energy. Although matter can be neither created nor destroyed, it can be converted from one state to another (i.e., from solid to liquid, then to gas, and reversed). The overwhelming experience of experimental physics confirms this first law to be a fact.

However, this law has enormous implications regarding the origin of the matter in our universe.

Halton C. Arp (b. 1927), a professional astronomer who, in his early career, was Edwin Hubble's assistant, has earned the Helen B. Warner prize, the Newcomb Cleveland award, and the Alexander von Humboldt Senior Scientist Award. For years, Mr. Arp worked at the Mt. Palomar and Mt. Wilson observatories, where he developed his well-known catalog of peculiar galaxies (galaxies that are irregular in appearance because of interaction and merging).

Arp, as well as his contemporary astronomers, discovered through observations that galactic redshifts are possibly due to other mechanisms, such as atoms having variable mass. If this were true, this would mean that the universe is not expanding at all. If the universe is not expanding, then it contradicts the well-accepted theory of the big bang. Arp discovered, by taking photographs through huge telescopes, that many pairs of quasars have extremely high redshift values. Therefore, they must be receding from us at incredibly high speeds—and consequently be located at a great distance from us. He also discovered that these quasars were physically associated with galaxies that have low redshift and therefore must be at relatively close distances. These exciting observations led to the conclusion that scientific and astronomical assumptions about the age of the universe and its dynamical state cannot be correct. Arp became an outspoken critic of the big bang theory, causing much commotion in the astronomical community. Because of a serious effort to suppress this evidence, Dr. Arp was ousted from the astronomical community and denied telescope time. Scientific journals refused to publish his findings, and he was no longer invited to speak at conferences.

It is interesting to note that the Halton Arp incident closely resembles the Galileo affair that took place some four hundred years earlier. Galileo made a remarkable discovery; he proved without a shadow of a doubt that the sun was at the center of the solar system, not Earth. However, his discovery contradicted the views of the church, and Galileo was ordered to shut up, go home, and stay there, which he did. After Mr. Arp's remarkable discovery, which contradicted the big bang theory, he was also ordered to shut up, go home, and stay there.

However, Arp has never proclaimed that the big bang never occurred or that Hubble's law is wrong; he simply states that he made observations that contradict popular accepted theories. He confessed that "The Big Bang Theory is false - not because others or I claim it to be false - but because it has been scientifically falsified." Halton C. Arp is now at the Max Planck Institute in Munich, Germany. Occasionally he returns to the United States to give lectures and visit family.

Geoffrey Burbidge (1925–2010), a British astronomy professor at the University of California, San Diego, gave the following devastating summary of the antiscientific conduct of the astrophysical establishment: "The existence of a class of objects which have red shifts not largely due to the cosmic expansion was not predicted either in the hot big bang cosmology or in QSSC (Quasi-Steady State Cosmology). How is this phenomenon dealt with in each hypothesis? As far as the big bang model is concerned, its supporters are in complete denial. They never mention the observational evidence, do not allow observers who would like to report such evidence any opportunity to do this in cosmology conferences, argue against its publication, and if forced to comment on the data, simply argue that they are wrong."

The cosmic microwave background, or CMB, a uniform radio fog surrounding Earth, has been artificially interpreted as an image of the primordial fireball. It is no such thing. Despite persistently flawed instrumentation that could not possibly produce sufficiently accurate data, the COBE satellite was nevertheless credited with measuring the most perfect blackbody ever recorded in the history of science. The radio fog surrounding us consists only of ambient starlight reflecting local structure and the equilibrium temperature of space. It cannot logically be connected to an expanding universe or primeval fireball. The statement that the big bang theory explains the observed microwave background is a distortion of the meaning of words, according to Professor Burbidge.

Thomas Van Flandern (1940–2009) was an American astronomer and author specializing in celestial mechanics. In his career as a professional scientist, he became famous as an outspoken proponent of nonconventional views related to astronomy, physics, and extraterrestrial life. In his paper *The Top 30 Problems with the Big Bang*, he gives an overview of some of today's problems in cosmology:

> The Big Bang ... no longer makes testable predictions wherein proponents agree that a failure would falsify the hypothesis. Instead, the theory is continually amended to account for all new, unexpected discoveries.
>
> Indeed, many young scientists now think of this as a normal process in science! They forget, or were never taught, that a model has value only when it can predict new things that differentiate the model from chance and from other models before the new things are discovered.
>
> Perhaps never in the history of science has so much quality evidence accumulated against a model so widely accepted within a field. Even the most basic elements of the theory, the expansion of the universe and the fireball remnant radiation, remain interpretations with credible alternative explanations.
>
> One must wonder why, in this circumstance, four good alternative models are not even being comparatively discussed by most astronomers. [One of these models is quasi-steady-state cosmology (QSSC), proposed in 1993 by Hoyle, Burbidge, and Narlikar.]
>
> Big Bang Cosmology, which is built on general relativity theory, is forced to use a number of adjustable parameters and ad-hoc assumptions to agree with observation, such as inflation. The assumption that most of the mass of the universe must consist of "dark matter," a kind of matter that cannot be detected, but nevertheless must exist, for the sole reason that Big Bang theory requires it, and now the latest fad, "dark energy."
>
> Two of the three vaunted "predictions" of Big Bang theory - the light element abundances and the temperature of the microwave background are actually retro dictions meaning that Big Bang theory failed to predict them

quantitatively correctly and then the data was adjusted to fit the observational evidence. [In other words, the facts were manipulated to fit the theory, so to speak.]

The third, the Hubble expansion, is entirely a figment of the imagination, as veteran astronomer Halton Arp has pointed out for decades. There are ample examples of high-red shift quasars that are physically connected to low-red shift galaxies, and there is evidence that red shift is quantized. However, astronomy has failed to self-correct, and the only acknowledgement Arp received from the scientific establishment was to be largely (though not completely) banned from publication in scientific journals or from speaking at conferences, and to be denied telescope time.

Although he was credited with the theory of an expanding universe, the great Edwin Hubble was also the first to admit that the evidence for expansion was questionable. He claimed universal expansion is not supported by observation but is actually contradicted by it. The original data indicating expansion were found dubious, then abandoned, but never replaced. It was Lemaitre, not Hubble, who originated the Hubble law, and Hubble spent the rest of his life trying to explain this to people. It came from nowhere and, observationally, it is not going anywhere. The fictitious emergence of universal expansion came from incorrect data. In 1947 Dr. Hubble admitted, "It seems likely that redshifts may not be due to an expanding universe, and much of the speculation on the structure of the universe may require re-examination."

In May 2004, a group of about thirty concerned scientists published an open letter to the global scientific community in *New Scientist* in which they protested the stranglehold of the big bang theory on cosmological research and funding. The letter was placed on the Internet at www.s8int.com/bigbang5.html and rapidly attracted wide attention. It is supported by signatories representing scientists, astronomers, and researchers from around the world of disparate backgrounds and has led to the formation of a loose association known as the Alternative Cosmology Group. The letter questions a fundamental belief—the belief in the so-called big bang theory. Therefore, it will be interesting to monitor the response of that community. Already, the first line of defense—censorship—has held. The journal *Nature* rejected the letter for publication. *New Scientist*, the more popular magazine, finally published the letter under the title "Bucking the Big Bang" on May 22, 2004.

Theology and science have been at odds ever since George Lemaitre introduced his cosmic egg theory. Theology asserts that if there is a cosmic egg, then there must be a cosmic chicken as well; whereas science believes a cosmic chicken is not necessary. The good news is that the two sides can be compatible. The bad news is that resolution between the two will most likely not happen any time in the near future. The late famous atheist astronomer Carl Sagan boldly proclaimed in 1996 that he saw no evidence of a God to believe in, which confirms that science and theology have become two parallel lines, never intersecting and yet remaining in constant conflict. Carl Sagan's belief that the cosmos is all there is, all there was, and all there ever will be cannot be taken seriously.

The problem is that one or both groups are wrong. Either bad science or bad religion is causing a great deal of problems. The scientific discoveries of the last century do not declare that God does not exist. In fact, they offer an understanding of how God created the universe. It is the Christian community's insistence on a young Earth creation that has caused the scientists to believe that the Bible is in error and that therefore the God of the Bible cannot be real.

However, reconciliation is realistic in the science/Creation conflict. As soon as both sides are willing to listen to each other, then the healing process can begin. As long as both parties realize that "It is just as foolish for science to ignore the truths of the Bible, as it is for theology to disregard discoveries in science."

Theology has always insisted that our universe was created by the word of God. Who then is God? Who then created God? Did God have a first cause? The answer is, God is an uncreated eternal being without a beginning or end, not bound by, or restricted to the laws of nature, time, or space. God is of a dimension the human mind is incapable of comprehending, therefore God cannot be scientifically or biologically explained. The Bible tells us that God is eternal, without beginning or end. He has always existed. This makes no sense from our limited understanding until we accept that time and space are not part of God's limitations. God created both of them in the beginning. Our limitations are not God's limitations. As such, God is not physically perceivable unless He purposely discloses Himself. He has done just that. The creation of our universe declares His glory, His majesty, and His power, and He has given us His word. God is the only entity that is capable of performing "supernatural acts," or miracles. Scriptures tell us fifty-eight times that God is "Almighty," which means He is supernatural, and scripture tells us that God was, is, and will be, meaning that God has always been here, is here today, and will be here for eternity. God is omnipotence, omniscience, and omnipresence. Theology believes that God is in fact the Intelligent Designer, Supreme Architect, and Creator of our universe—an incredible miracle indeed! The Bible not only proclaims such a Creator but also proves His existence by demonstrating that the biblical text came from an extra dimensional, transcendent, supernatural source beyond time and space. One day, scripture promises, everything about God and His character will be revealed to us.

"Theologians generally are delighted with the proof that the universe had a beginning, but astronomers are curiously upset. It turns out that the scientist behaves the way the rest of us do when our beliefs are in conflict with the evidence," according to Robert Jastrow.

The atheist's model of the origin of the universe begins with an even more impressive miracle—the appearance of all the matter in the universe from nothing, by no one, and for no reason. The atheist does not believe in the inspirational First Cause we call God. The atheist has no natural explanation, and neither does he have a supernatural explanation, for the origin of space-time and matter. Consequently, the atheistic scenario on the origin of the universe defies all rules of logic. He begins his model for the universe with an incredible miracle, well knowing that only the God of the Bible is capable of performing such a miracle, or supernatural event; but for the atheist, there is no room for a God—any God!

> In the beginning was the Word, and the Word was with God. (John 1:1)
>
> And, Thou, Lord, in the beginning didst lay the foundation of the earth, and the heavens are the works of thy hands. (John 1:1)
>
> For thus saith Jehovah that created the heavens, the God that formed the earth and made it, that established it and created it not a waste, that formed it to be inhabited: I am Jehovah; and there is none else. (Isaiah 45:18)

In 1991, the newly constructed Hubble telescope, observing an area of space equal to a grain of salt held at arm's length, discovered that there were an estimated 130 billion more galaxies in our universe, each containing hundreds of billions of stars. It seems like every time humanity invents and builds improved telescopes, God reveals more of His wondrous creation. It is as though God is saying, "Go build a bigger one, and I will show you more—things you have never seen or even imagined before!" Every time we make new discoveries about the universe, we find it is much bigger than we had imagined before. It seems that our universe is expanding after all; or at least our knowledge of our universe is expanding. Still, nobody really has any idea of how big the universe actually is, or where it ends—if there is an end to it. Therefore, astronomers and scientists always talk about the "known" universe. Where does the universe end, and what is there beyond the end of the universe? Nothing? No, there is God. Scriptures tell us that God can hold the universe in the palm of his hand.

Brian Greene, a world-renowned physicist and astronomer, states in his book *The Fabric of the Cosmos: Space, Time, and the Texture of Reality*,

> There's no way that scientists can ever rule out religion, or even have anything significant to say about the abstract idea of a divine Creator. Every piece of data that we have indicates that the universe operates according to unchanging, immutable laws that don't allow for the whimsy or divine choice to all of a sudden change things in a manner that those laws wouldn't have allowed to happen on their own.

The universe is incredibly wondrous, incredibly beautiful, and it fills me with a sense that there is some underlying explanation that we have yet to fully understand," Then he said, "If someone wants to place the word God on those collections of words, it's okay with me.

We can visually observe all the marvels of modern architecture, including the tallest buildings on the face of the planet, for example the Empire State Building in New York (381 m), Burj Khalifa in Dubai (828 m), Petronas Twin Towers in Kuala Lumpur (452 m), and Taipei 101 in Taiwan (509 m) among thousands of others. Every time we observe and admire these marvels of architecture, we realize they did not come into being by themselves, from nothing, without a purpose. Everyone realizes that it takes an individual with an idea—more likely a group of individuals—to form a plan or conception to get the ball rolling on such projects. The ultimate purpose is to provide retail space, office space, and residential units for rent or resale and realize a profit in the end. After all the preliminaries are completed, feasibility studies are satisfactory, and budgets meet approval, then a meeting is called and the final decision is made either to proceed with the project or to abandon it. Once the decision is made to continue, a team of very intelligent people—groups of partners, architects, engineers, contractors, and, of course, a consortium of bankers—are selected and recruited to make these projects possible. The blueprints are drawn up, studied, and approved. The architects design the visual appearance. The engineers makes sure everything is safe and sound; they makes sure the walls, columns, and pillars on the main floor have the integrity to support the floors above, realizing that a simple two-by-four is not going to do the trick. Soil tests are conducted, political issues are negotiated, and the contractors assemble their work crews. They then go to work and put everything together precisely, according to the plans and designs. Millions of dollars change hands even before the first line is drawn on paper or the first nail is hammered.

It is obvious that these projects are the results of a very specific procedure. Although as bystanders we most likely never have the opportunity to meet the source of intelligence, such as the architects and engineers, personally, we still realize their existence. Without them, these projects could never materialize. It would only be foolish to believe that these projects would come into existence by themselves from nothing, without designers, plans, or any source of intelligence. There is no one on the face of Earth, either bartender or astronomical scientist that would ever entertain such a thought. Yet some scientists are trying to convince us that the universe, which is much more complicated and complex than even the most sophisticated piece of architecture on the face of Earth, accidently "banged" into existence by itself without any consideration or planning and for no reason.

When we observe the universe, either with the naked eye or through a telescope, we cannot help but notice the existence of the very same principle and procedure, the existence of a very intricate design, where every detail is precisely organized, harmonized, engineered, and fine-tuned, meaning that there must be a definite source of intelligence behind the scenes. Because no design can exist without a designer, nor a creation without a creator, this proves beyond any reasonable doubt the existence of a supernatural entity of a much greater dimension than the universe itself, not bound by or restricted to the laws of the universe. This could be no one else than the Supreme Architect, Designer, Engineer, Creator, Almighty God himself. Even though we may not meet God face-to-face in physical form (not yet anyway), this does not mean that He does not exist or is not there. "Of old hast thou laid the foundation of the earth: and the heavens *are* the work of thy hands? They shall perish, but thou shalt endure: yea, all of them shall wax old like a garment; as a vesture shalt thou change them, and they shall be changed" (Psalms 102:25–26).

In order to arrive at a logical conclusion of how the universe originated, one must have an understanding of what "logic" really means. Merriam-Webster's dictionary describes logic as "the study of reasonable thinking." Logic means arriving at a conclusion based on facts. A fact is something that is known and proved true, without reasonable doubt. For example, the statement "Humans are mortal" is considered to be a fact, as is the statement "George Bush is a human."

Considering the two statements, we can logically conclude that George Bush is mortal. We can logically arrive at this conclusion because the two statements are interrelated, meaning that one fact is dependent on the other or affected by the other. We can arrive at the logical conclusion that George Bush is going to die because George Bush is human and humans are mortal. Now let us consider the two following facts: "Apes have hair" and "George Bush has hair." Based on the two facts, we cannot logically conclude that George Bush is an ape, because the two facts are not interrelated. The two statements are two independent facts; the first declares that apes have hair, and the second declares that George Bush has hair also. However, so do lions, cats, and dogs, and they are no apes either.

We have examined three different possibilities of the universe's origin. The first is the static or steady-state universe, meaning the universe has always been here, and had no beginning, no first cause, and no plausible reason to be here. Logically, this is, of course, not possible. If there were no first cause, then there would be absolutely no reason or purpose for the universe to be here.

"Ex nihilo" is a Latin term that means "from out of nothing." The Romans used the phrase "ex nihilo nihil fit," meaning "Nothing comes from nothing." This was an idea presented by St. Augustine that later became important to Church doctrine. It represented Augustine's philosophical explanation of how God created everything from nothing. Interestingly, science also implies that the universe also originated from nothing as well. Exactly what is nothing, and what does "nothing" really mean? Is it just an empty void or an empty space? No, an empty void or an empty space is something indeed; it consists of an empty void or an empty space. No one has ever been able to offer a credible explanation of what "nothing" really means, other than "not anything" or "lack of something." "Nothing" refers to a condition of nonexistence; it is an indefinite pronoun indicating that there is not anything or not a single part of a thing. Nothingness is of another dimension, just like Almighty God Himself, a dimension that our limited human mind is not capable of comprehending. If there ever was a time when nothing existed, there would be nothing in existence today, because nothing produces nothing other than nothingness! There is no scientific data that indicates matter has the ability to create itself. Logically, we must conclude that the universe had a nonmaterial origin. Atheist science cannot explain the order or design that is characteristic of our universe. The theory of a static or steady-state universe demands rejection because of the following argument: the facts that "Ex nihilo nihil fit" (nothing comes from nothing) and the universe does exist logically lead us to conclude that the universe had a beginning. Indeed, it came from somewhere. Regardless of these facts, some scientists completely deny logic and continue to build theories on a static universe.

The second account of the origin of the universe is the big bang theory. Now, this is where things start to make a little more sense. The big bang theory at least acknowledges the fact that there was an actual beginning to the universe; however, it is this particular beginning that puzzles everybody, including scientists, astronomers, and theologians. The big bang theory tries to explain that all of the matter and energy in the entire universe started as a point of infinite density and temperature, labeled a singularity. A singularity defies and contradicts all laws of physics. It is logically not possible that all the mass and energy in the entire universe could be condensed into a tiny little ball much smaller than the size of a marble. Some scientists suggest that the entire universe, including all its matter and energy, the sun, Earth, and the trillions of other stars and billions of other galaxies, was once condensed into this tiny, very hot something, and then, in one millionth of one millionth of one millionth of a second, it expanded to its present state. A rapid expansion of this magnitude is, of course, not possible, because it suggests that matter in the universe moved millions of times faster than the speed of light, even though scientists insist that nothing can move faster than the speed of light. Despite the fact that nobody really has a plausible explanation of what a singularity really is, or where it came from, some astronomers and scientists still insist that other singularities probably exist in the cores of black holes.

Although it survives as the most accepted theory about the origin of the universe today, and is supposedly supported by substantial observational evidence, opposing scientists insist that the big bang is based on a false principle.

Marilyn Vos Savant published an interesting article in *Parade* magazine on February 4, 1996, which stated, "I think that if it had been a religion that first maintained the notion that all the matter in the entire universe had once been contained in an area smaller than the point of a pin, scientists probably would have laughed at the idea." In fact, the Bible does say something remarkably similar, and yes, it has been ridiculed and laughed at by many. "In the beginning God created the heavens and the earth" (Genesis 1:1). "That alone stretcheth out the heavens, And treadeth upon the waves of the sea" (Job 9:8). "By faith we understand that the worlds have been framed by the word of God, so that what is seen hath not been made out of things which appear" (Hebrews 11:3).

The big bang theory has been around for such a long time and has been discussed so often that scientists now simply accept the theory as fact, hereby changing the fundamental meaning of the word "theory" itself. To the common person, the word "theory" still has the original meaning: "A guess, possibility, or assumption." However, to the scientist, the meaning of the word has changed to "A belief that has been generally accepted as a result of actual experimentation and/or observation." However, the big bang cannot provide any evidence for the theory to be classified as proven fact. Obviously, the big bang theory is disintegrating from within. The assumptions of the standard model are illogical, in direct conflict with observation, and supported only by great uncertainty. Facts are continuously amended to fit the theory instead of the theory being amended to fit the facts. We have no good reason to believe what we are expected to believe. No logical philosophy should be built upon such foundations. Yet when we watch astronomy programs on the History or Discovery Channels, astronomers usually start with a declaration that the origin of the universe remains to be a profound mystery, then they continue to talk about the big bang as though it is a fact and as if it happened just yesterday, just as if they were present, on location, when it actually happened. Ironically, hardly ever is there any mention of a God-created universe. I truly believe that some people can actually be educated to the point where the basic laws of logic and common sense no longer have any meaning or value. I also believe the big bang theory provides a prime example of this, because the entire big bang theory is based on assumptions, which are based on another assumptions; there is no actual conclusive proof of anything. Moreover, just like the steady-state theory, for a universe to establish itself from absolutely nothing for no reason, without a purpose, is simply nothing more than just a dream or a scientific fairy tale. The basic laws of logic and common sense simply do not permit these absurd stories to be accepted as truth. There is just as much, if not more, evidence disproving the big bang theory as there is supporting it. This brings us to the only logical and reasonable conclusion—that the biblical description of Divine Creation is the only alternative.

NASA literally spent billions of dollars to send probes into outer space to explore the origin of our universe. However, all they really needed to do was invest about fifteen dollars in a Bible, and there they would find the answers they were looking for.

Science and theology do have something in common; they both share a quest for the ultimate truth, and it would be mere foolishness for a theologian to deny and ignore the advances of scientific studies and discoveries, just as it would be for a scientist to deny and ignore the truths of the Holy Bible.

After the creation of humanity, God selected and inspired some people to write down things about humanity and things about God Himself in a book called the Holy Bible. The contents of the Holy Bible over the last centuries have been much in dispute and much debated; however, they have never been proven wrong or incorrect. The Bible clearly states that Almighty God is the Supernatural Creator; that He is the only one and that there is no other beside Him. Every phenomenon or occurrence in science for which there is no scientific explanation, which defies the rules of the universe or the laws of nature, is considered an act of a supernatural being; there is simply no other reasonable or acceptable explanation. If the aforementioned singularity really did exist, defying all the rules of the universe and the laws of nature, obviously it must have been a supernatural event. Only the God of the

Bible is supernatural and is not bound by or restricted to the laws of the universe and nature. Therefore, fundamentally, scientists do in certain ways acknowledge a supernatural creator; however, they do seem to have a problem calling it by its proper name. Perhaps they find this is too religious. The scientific explanation of a singularity sounds a lot more like a story that belongs on the cover of the *National Enquirer* or some science fiction magazine than one that belongs in a scientific journal.

How, then, did God create the universe? According to scripture, God created the universe, humanity and all living things by "His Word." In the first chapter of Genesis, the words "and God said" are used nine times, and every time God said something, something happened, something was created, something come into being from nothing. Consequently, the word of God was the First Cause; the word of God was the action that caused the reaction of the universe's existence—an incredible miracle indeed—meaning that the universe had an actual beginning and a purpose as well. It is obvious that humanity is God's sole purpose for the creation of the universe. Genesis seems to be vague and lacks many of the chemical and biological details. The answer to the question of how the universe came into being is actually simple; there is not a single soul on the face of Earth that has a precise logical explanation. The origin of the universe still is a profound mystery, and it will probably remain so for many years to come, or probably for as long as humanity exists. Scientists have all sorts of theories and ideas, but their theories are continually proven inaccurate or wrong by opposing scientists.

The fact does remain that God did create the heavens, the earth, and all living species upon the earth. This is a fundamental truth that must be accepted by the scientific community as such, in order to get somewhere in the quest for the ultimate truth. Moreover, is it actually important to be aware of every minute detail and criterion? Probably not. If it were, God would have revealed many more details that are relevant. Einstein once said, "I want to know how God created this world. I am not interested in this or that phenomenon, in the spectrum of this or that element. I want to know his thoughts, the rest are details." Although Einstein never acknowledged a personal God, he did concede to a supernatural creator; he even referred to this creator as "God." Einstein was trying to convince scientists that science is the discovery of how and when Almighty God created all things, including the universe. The theory of singularity is just another scientific fairytale; there is no plausible explanation for its existence, other than through a supernatural act of Almighty God. However, who knows, perhaps there was sort of a bang when God created our universe; the Bible does not say.

The book of Genesis describes eight acts of creation within a six-day framework, followed by a day of rest. Each of the first three days represents an act of division: day one divides the darkness from light, day two divides the waters from the skies, and day three divides the sea from the land and places vegetation on Earth. In each of the following three days, these divisions are populated. Day four populates what was created on day one, and heavenly bodies are placed in the darkness; day five populates what was created on day two, and fish and birds are placed in the seas and skies; finally, day six populates what was created on day three, and animals are placed on the land. Then God said, "Let us make human beings in our image," and he created humanity. This six-day structure is symmetrically bracketed: on day zero primeval chaos reigns, and on day six there is organization and order in the cosmos.

Genesis 2 provides a simpler linear narrative and starts out with God giving His approval of the universe as "very good" and placing humanity at the highest priority of the creation order. Then things deteriorate from the initial state of being "very good": Adam and Eve eat the fruit of the Tree of Life in disobedience of the divine command. Only ten generations later, in the time of Noah, Earth has become so corrupted that God decides to return it to the waters of chaos, sparing only one righteous man and his family in order to continue humanity.

However, the entire Creation narrative of Genesis is described in just one single page of the Bible, and therefore it remains vague and cannot describe a precise sequence of events of how the universe originated. Because of the lack of details, different groups of Creationists evolved, each proposing their own interpretation of the book of Genesis. What is Creationism? What is creation science? What is intelligent design?

Creationism, creation science, and intelligent design are three religious concepts of creation trying to explain the origins of the universe. Sometimes it is difficult to distinguish among these theories.

Creationism is the belief that God has created the universe and humankind, typically as presented in the Bible's book of Genesis. The book clearly affirms that the matter in our universe arose from an act of a transcendent creator at a finite point in time. However, the materialistic worldview has no room for a creator. Therefore, an alternate explanation needed to be found.

Creation science attempts to provide evidence that the world was created by God by proving evolution wrong and then offering interpretations of scientific data that prove the Creation account in Genesis to be correct.

Intelligent design theorists similarly offer a theory of God's role in the Creation, arguing that the very complexity and organization of the world—and the failure of science to explain it all—makes God's intervention the only reasonable explanation. Intelligent design theorists do not typically rely on Genesis; instead they attempt to provide evidence of God's role in creation in their observations of the world.

How do we know that Almighty God did create our universe? The simple answer is that He told us so. Almighty God manifested himself by inspiring a number of privileged men, forty-four of them, to record His instructions for the rest of his newly created humanity in a series of books called the Bible or the Word of God. The Bible is the first book ever written and is today still the best-selling book of all times. The Bible is the story of humankind from the very beginning (Genesis) to the very end (Revelation). I strongly recommend that anyone who has any reservations about the authenticity and reliability of the Holy Scriptures read Dr. Grant Jeffrey's fascinating bestselling book *The Signature of God*. In his book, Dr. Jeffrey examines exceptional scientific, medical, historic, archeological, and prophetic discoveries that conclusively prove beyond any reasonable doubt that the Holy Bible is truly the inspired Word of God. The Bible also tells us that God did not hide Himself from humanity but rather declared his presence by the work of his hands. God's hand is obviously evident in creation, and it is simply foolish to deny God after observing a great witness like the universe.

The number seven denotes the number of God's divine perfection and completion. It is embedded in the text of Genesis 1 (but not in Genesis 2) in a number of ways, besides the obvious seven-day Creation framework: The word "God" occurs thirty-five times (7 × 5) and "Earth" twenty-one times (7 × 3). The phrases "and it was so" and "God saw that it was good" occur seven times each. The first sentence of Genesis 1:1 contains seven Hebrew words composed of twenty-eight Hebrew letters (7 × 4), and the second sentence contains fourteen words (7 × 2), while the verses about the seventh day (Genesis 2:1–3) contain thirty-five words (7 × 5) in total. In the English translation, the very first sentence, "In the beginning," totals fourteen letters composed of seven different letters. It is also interesting to note that humanity will exist for a period of seven thousand years and the total population of Earth will shortly reach seven billion people.

The first five books of the Old Testament, called the Torah, were written by Moses, who was literally handpicked by Almighty God. Moses was born at a time when the Hebrew population was enslaved in Egypt. They were increasing in numbers. Consequently, an unnamed Egyptian pharaoh commanded that all male Hebrew children born be killed by drowning in the Nile River, in order to control the population of the slaves. Jochebed, the wife of the Levite Amram, bore a son, named him Moses, and kept him concealed for three months to avoid the murder of her newborn. When she could no longer keep the boy hidden, rather than deliver him to be killed, she set him adrift on the Nile River in a small basket of bulrushes coated in pitch. Moses' sister Miriam watched the little craft float down the river until it was spotted by Pharaoh's daughter, who was bathing in the river with her maidens, and she rescued Moses. He was adopted and raised at Pharaoh's royal palace, obviously getting the best of everything, including the best education, certainly qualifying Moses for the task of recording the Torah.

When Moses was a young man, in a fit of rage, he killed an Egyptian slave master who was beating a Hebrew slave. Moses fled and escaped to the land of Median. After about forty years, Almighty God manifested Himself in the form of a burning bush on the side of Mount Sinai and instructed Moses to return to Egypt. After his return to Egypt, Moses liberated the Hebrews from their bondage in Egypt, wandered in the wilderness for forty years, and led them to freedom in the Promised Land. By the inspiration of God, Moses recorded their story in the book of Exodus, and he authored Genesis, the Creation narrative, as well. On Mount Sinai, Moses also received the Ten Commandments as well as the law that would govern the Jews as well as the Gentiles for centuries to come.

> In the beginning God created the Heavens and the Earth.
> (Genesis 1:1)

> … who is the image of the invisible God, the firstborn of all creation; for in him were all things created, in the heavens and upon the earth, things visible and things invisible, whether thrones or dominions or principalities or powers; all things have been created through him, and unto him; and he is before all things, and in him all things consist.
> (Colossians 1:15–17)

> Worthy art thou, our Lord and our God, to receive the glory and the honor and the power: for thou didst create all things, and because of thy will they were, and were created.
> (Revelation 4:11)

Genesis 1:1, "In the beginning God created the Heavens and the Earth," could very well be considered a general Creation statement, followed by details that are somewhat more explicit regarding the subsequent days.

The narrative of Genesis could indeed be interpreted in several different ways and could be viewed from different perspectives. For example, is it not possible that Genesis's "Creation week" could be a description of the entire duration of the universe, from the very beginning to the very end of humanity, after the millennium? Is it not possible that day seven, the day God rested, has not yet occurred?

Literalists believe that every day of the Creation week represents a precise twenty-four-hour period. Our current system of measuring time is based on the rotation of Earth around its axis in a twenty-four-hour period, called a day; and Earth's orbit around our sun in 365¼ days is called a year. Genesis 1:14–18 clearly refers to the placement of our sun and moon in our solar system on the fourth day, so what system of time measurement could possibly be subscribed to the first three days? Is it not more likely that every Creation day could be considered a period of time, but not necessarily a twenty-four-hour day? When God said, "Let the earth bring forth plants," He actually authorized to the earth to produce plant life, probably from the very resources of the earth itself by the natural laws God had established at the origin of the universe. "Let the earth bring forth plants" could indeed involve a long time—a lot more time than just one single twenty-four-hour day!

We must also consider the fact that the Holy Bible consists of three main themes: history, prophecy, and poetry. Is it not reasonable to accept that the Creation narrative could be considered a poetic rendering of the Creation story, rather than a historic rendering? How can one explain, for example, the existence of Earth, already vegetated, on the third day, prior to the existence of our sun and moon—or our solar system, for that matter? Did our solar system not form simultaneously according to geological and astronomical discoveries? Does vegetation not require energy provided by our sun to exist, grow, and flourish?

The integrity of the scientist also plays an important role in the discovery of the origin of the universe. Is he an atheist, or is he a believer in God? His passions will certainly affect his theories and hypotheses. As an atheist, he will most likely never have any consideration for the involvement of God or an intelligent designer in his theories. He is denying the fundamental truth; exalting himself above God, so to speak. He will likely continue to pursue absurd and bizarre ideas regarding the origin, sometimes even resorting to science fiction.

We must also consider the Christian Scientist. Is he a scientist that happens to believe in God, or is he a believer in God that happens to be a scientist? Does he propose a scientific theory and then try to fit God in there somewhere, or is he a believer in God who already has a preliminary idea of how God thinks, and then builds his theories on the analysis of the details? The general public, and the religious community in particular, have become highly suspicious of modern science. Scientists accomplish a great many remarkable things, but despite their achievements, many are concerned that science has lost its way. No longer seen as trying to understand the work of God, it is perceived as pushing an anti-God agenda. A few outspoken scientists—Richard Dawkins, Stephan Hawking, Carl Sagan, and Stephan J Gould, to name a few—are openly hostile toward the Christian faith. Though this may not be the norm, it does increase the distrust of the public. It is actually sad to see that people of such intellectual significance can lower themselves to the point where their theories belong nowhere other than the funny papers instead of scientific journals. How appropriate is the verse "Professing themselves to be wise, they became fools" in Romans 1:22?

However, I sincerely do believe that a vast number of atheist or agnostic astronomers and scientists are redirecting their focus and paying more attention to the words of Einstein: "I want to know how God created this world. I am not interested in this or that phenomenon, in the spectrum of this or that element. I want to know his thoughts, the rest are details." Jim Holt of the *Wall Street Journal* wrote in 1997, "I was reminded of this a few months ago when I saw a survey in the journal *Nature*. It revealed that 40 percent of American physicists, biologists, and mathematicians believe in God—and not just some metaphysical abstraction, but a deity who takes an active interest in our affairs and hears our prayers: the God of Abraham, Isaac, and Jacob."

A vast number of astronomers and scientists have concluded that a universe cannot exist without some sort of supernatural plan or supernatural agency or supernatural source of intellect. Here is what some have to say, regardless of whether they are atheists, agnostics, or believers in God:

Frank Tipler (b 1947), is a professor of mathematics, physics and cosmology at Tulane University. "From the perspective of the latest physical theories, Christianity is not a mere religion, but an experimentally testable science."

Wernher von Braun (1912–1977) was a German American pioneer rocket engineer. "I find it as difficult to understand a scientist who does not acknowledge the presence of a superior rationality behind the existence of the universe as it is to comprehend a theologian who would deny the advances of science."

John O'Keefe (1916–2000), a practicing Roman Catholic, was a planetary scientist with NASA from 1968 to 1995. "We are, by astronomical standards, a pampered, cosseted, cherished group of creatures … If the universe had not been made with the most exacting precision we could never have come into existence. It is my view that these circumstances indicate the universe was created for man to live in."

Alan Sandage (b 1926) is best known for determining the first reasonable accurate value for the age of the universe. He was a graduate student working with Edwin Hubble and continued Hubble's work after his sudden death in 1953. "I find it improbable that such order came out of chaos. There has to be some organizing principle. God to me is a mystery but so is the explanation for a miracle of existence, Why is there something instead of nothing?"

Stephen Hawking (born on January 8, 1942) has become one of the world's most famous theoretical physicists. His scientific career spans well over forty years. His books and public appearances have made him an academic celebrity, and in 2009, he was awarded the Presidential Medal of Freedom, the highest civilian award in the United States. Hawking takes an agnostic attitude toward religious matters. He has repeatedly used the word "God" to illustrate points made in his books and public speeches. Hawking has stated that he is "not religious in the normal sense" and that he believes "the universe is governed by the laws of science. These laws may have been decreed by God, but God does not intervene to break the laws." Famous British author and composer Anthony Burgess commented that in *A Brief History of Time* Hawking brings in God as a useful metaphor, however is an atheist. Although a self-proclaimed atheist, Hawking does concede the following: "The odds against a universe like ours emerging out of something like the big bang is enormous ... I think, clearly, that there are religious implications whenever you start to discuss the origins of the universe. There must be religious overtones. But I think most scientists prefer to shy away from the religious side of it." He also commented:

> The laws of science, as we know them at present, contain many fundamental numbers, like the size of the electrical charge of the electron and the ratio of the masses of the proton and the electron. The remarkable fact is that the values of these numbers seem to have been very finely adjusted to make possible the development of life.
>
> It would be very difficult to explain why the universe should have begun in just this way, except as the act of a God who intended to create beings like us.
>
> Then we shall ... be able to take part in the discussion of the question of why it is that we and the universe exist. If we find the answer to that, it would be the ultimate triumph of human reason—for then we would know the mind of God.

Arno Penzias (b. 1933), American physicist and 1978 Nobel Prize winner, said, "Astronomy leads us to a unique event, a universe which was created out of nothing, one with the very delicate balance needed to provide exactly the conditions required to permit life, and one which has an underlying (one might say 'Supernatural') plan. The best data that we have are exactly what I would have predicted, had I nothing to go on but the five Books of Moses, the Psalms, the Bible as a whole."

Paul Davies (b. 1946) is an English physicist, writer, and broadcaster who is currently a professor at Arizona State University. He stated, "The laws which enable the universe to come into being spontaneously seem themselves to be the product of exceedingly ingenious design. If physics is the product of design, the universe must have a purpose, and the evidence of modern physics suggests strongly to me that the purpose includes us."

Tony Rothman (b. 1953), an American theoretical physicist and writer, said, "When confronted with the order and beauty of the universe and the strange coincidences of nature, it's very tempting to take the leap of faith from science into religion. I am sure many physicists want to. I only wish they would admit it."

Robert Jastrow (1925–2008) was an American astronomer and cosmologist. He joined NASA when it was formed in 1958 and served until his retirement in 1981. He was also a self-proclaimed agnostic. Jastrow stated, "For the scientist who has lived by his faith in the power of reason, the story ends like a bad dream. He has scaled the mountains of ignorance; he is about to conquer the highest peak; as he pulls himself over the final rock, he is greeted by a band of theologians who have been sitting there for centuries."

Here are some of the many Bible verses that relate to the origin of our universe:

> In the beginning was the Word, and the Word was with God. (John 1:1)
>
> Who hath measured the waters in the hollow of his hand, and meted out heaven with the span, and comprehended the dust of the earth in a measure, and weighed the mountains in scales, and the hills in a balance? Who hath directed the Spirit of Jehovah, or being his counselor hath taught him? With whom took he counsel, and who instructed him, and taught him in the path of justice, and taught him knowledge, and showed to him the way of understanding? Behold, the nations are as a drop of a bucket, and are accounted as the small dust of the balance: Behold, he taketh up the isles as a very little thing. (Isaiah 40:12–15)
>
> Thus saith God Jehovah, he that created the heavens, and stretched them forth; he that spread abroad the earth and that which cometh out of it; he that giveth breath unto the people upon it, and spirit to them that walk therein. (Isaiah 42:5)

I have made the earth, and created man upon it: I, even my hands, have stretched out the heavens; and all their host have I commanded. (Isaiah 45:18)

For thus saith Jehovah that created the heavens, the God that formed the earth and made it, that established it and created it not a waste, that formed it to be inhabited: I am Jehovah; and there is none else. (Isaiah 45:18)

Where wast thou when I laid the foundations of the earth? Declare, if thou hast understanding. Who determined the measures thereof, if thou knowest? Or who stretched the line upon it? Whereupon were the foundations thereof fastened? Or who laid the corner-stone thereof? (Job 38:4–6)

The heavens are thine, the earth also is thine: The world and the fullness thereof, thou hast founded them. (Psalm 89:11)

Before the mountains were brought forth, Or ever thou hadst formed the earth and the world, Even from everlasting to everlasting, thou art God. (Psalm 90:2)

Of old didst thou lay the foundation of the earth; And the heavens are the work of thy hands. (Psalm 102:25)

The God that made the world and all things therein, he, being Lord of heaven and earth, dwelleth not in temples made with hands. (Acts 17:24)

For the invisible things of him since the creation of the world are clearly seen, being perceived through the things that are made, *even* his everlasting power and divinity; that they may be without excuse. (Romans 1:20)

And, Thou, Lord, in the beginning didst lay the foundation of the earth, And the heavens are the works of thy hands. (Hebrews 1:10)

By faith we understand that the worlds have been framed by the word of God, so that what is seen hath not been made out of things which appear. (Hebrews 11:3)

Worthy art thou, our Lord and our God, to receive the glory and the honor and the power: for thou didst create all things, and because of thy will they were, and were created. (Revelation 4:11)

It is very likely that mankind will never discover the precise details of how the universe actually came into being. The subject is expected to remain a profound mystery for future generations. Science so far has produced nothing more than fairy tales. The Bible states that Almighty God did design and create the universe, but it has nothing further to say about the precise particulars. The Bible does not describe one single detail concerning the biology or chemistry of the universe. I believe that astronomers and scientists should redirect their focus and start to explore how God created the heavens and the earth, rather than pursuing bizarre and impossible dreams and fantasies. It is not possible for a universe to design and create itself from nothing for no reason.

Evolution Controversy

Darwin's Theory of Evolution is the widely accepted theory that all life on Earth originated as the result of a random, natural process, where life forms gradually developed from one single common ancestor, expanding and evolving to escalate to the vast variety of organisms we know today. Ultimately, it tries to provide an explanation for the existence of humanity as an alternative to divine creation by Almighty God. The general theory of evolution is a philosophical viewpoint with atheistic implications. Creationism, on the other hand, is the belief that our universe and all life on Earth were created through a supernatural act of God, as told in the Bible, in the book of Genesis.

Merriam-Webster's Collegiate Dictionary gives the following definition of evolution: "a theory that the various types of animals and plants have their origin in other preexisting types and that the distinguishable differences are due to modifications in successive generations."

During the nineteenth century, it was a popular activity for people to visit museums exhibiting natural history collections. These museums would collect, catalogue, describe, and study vast collections of specimens from around the world. European naval expeditions employed naturalists and curators of these grand museums in search for live specimens of different varieties of life. Charles Darwin (1809–1882) was an English naturalist and geologist, best known for his contributions to evolutionary theory. Darwin's early interest in nature led him to neglect his medical education at the University of Edinburgh; instead, he helped to investigate marine invertebrates. Studies at the University of Cambridge encouraged his passion for natural science. He had a keen interest in the origin of life and signed on as a ship's naturalist on board the *HMS Beagle*, which sailed from Plymouth, England on December 27, 1831, dispensed to a five-year research expedition around the world. During his voyage, Darwin collected, studied, and observed an abundance of plants, animals, and fossils and paid a lot of special attention to the diverse forms of life along the west coast of South America and the neighboring Galapagos Islands. There he observed and researched species in an isolated environment. Darwin studied geological features, fossils, and animal behavior at the many stops the ship made as it sailed across the globe. He collected and returned with an enormous number of wildlife and fossil specimens and tried to solve the puzzle of how the vast variety of life forms had originated.

Upon his return to London in 1836, Darwin conducted thorough research of his notes and specimens he had collected from the many distant places. Out of this study grew several related theories: one, evolution did occur; two, evolutionary change was gradual, requiring thousands, perhaps millions, of years; three, the primary mechanism for evolution was a process he termed natural selection; and four, the millions of species alive today arose from a single original life form through a branching process he termed speciation. In 1839, Darwin published his findings in a journal known as *The Voyage of the Beagle*. In his journal, he described detailed scientific observations of biology, geology, and anthropology. Through further studies, he formulated the idea that each species had developed from ancestors with similar features. He described how the process he called natural selection could actually make this happen.

Darwin also noted that pollination in orchids occurs in a variety of different ways. He observed that organisms produce a lot more offspring than their environment can possibly support. Subsequently, he concluded that there must be a competitive struggle for survival where only one, or per perhaps a few individuals, can survive out of each generation. He also suggested that it was not chance alone that determined their survival. Instead, survival depends on the traits or characteristics of each individual, which can be either beneficial or detrimental to survival and reproduction. He suggested that well-adapted, or "fit," individuals are likely to leave more offspring than their less-well-adapted competitors. Darwin developed a theory that the imbalanced ability of individuals to survive and reproduce could be a cause for gradual changes in the population. Characteristics that help an organism survive and reproduce would accumulate over many generations. On the other hand, traits that hinder survival and reproduction would die out and disappear. Darwin used the term natural selection to describe this process.

In 1859, Darwin published his book *On the Origin of Species*, in which he first proposed his theory of natural selection, and it immediately gained a lot of attention in the scientific community. In addition, without Darwin's awareness, Gregor Johann Mendel (1822–1884), an Austrian Augustinian priest and scientist, experimented with the science of genetics, studying the inheritance of certain traits in pea plants in the garden of his monastery. Mendel discovered that the inheritance of these traits follows a particular set of laws, which later became known as Mendel's laws of Inheritance. The significance of Mendel's work was not recognized until the turn of the twentieth century, when independent rediscoveries of his studies formed the foundation of the modern science of genetics. Paleontology, or the study of fossils, became an important factor in the study of evolution and population genetics, and a global network of scientific research provided additional details into the mechanisms of evolution. Scientists now claim to have a better understanding of how new species can develop, and they claim to have observed the speciation process in laboratories and in the field. Evolution became the fundamental theory that biologists employ to understand life in general, and it is applied in fields like medicine, psychology, conservation biology, anthropology, forensics, agriculture, and other sociocultural applications. Speciation involves the process whereby offspring completely separate from parents and form a completely new species. However, today's understanding of genetics and the DNA code does not allow this to happen. Therefore, logically, this should be considered a scientific fairytale.

After many decades of intensive experimentation and research, the scientific community still does not have a precise idea, or even the faintest idea, of how life on Earth really did originate. The more science learns about the nature of the problems associated with life's origin, the more mysterious and complicated it seems to become. During the last decade, despite the fact that scientists have failed to provide conclusive evidence on how evolution allegedly does occur, evolutionary scientists have made a lot of effort to try to sell their theories and hypotheses to the general public using terms like "overwhelming evidence" and "astonishing evidence" in support of their position. For example, prominent atheist and evolutionary scientist Richard Dawkins claimed in an interview with journalist Bill Moyers that there is "massive evidence" for the theory of evolution, while in fact he failed to provide a logical explanation as to how life on Earth did originate, other than being "seeded" by unexplained aliens from some other distant part of the universe.

Evolution can be separated into two different divisions: "macroevolution" and "microevolution." Both divisions engage mostly the same biological principles and processes. Microevolution is the occurrence of small-scale changes in allele frequencies in a population over a few generations, also known as change below the species level. Macroevolution tries to explain major changes in all life forms over many generations through a process called evolutionary biology, or speciation, in which one species evolves into another one.

Arguments disputing macroevolution include the intelligent design arguments of irreducible complexity and specified complexity. Intelligent design proponents insist these complexities could possibly not be a result of an evolutionary process. However, neither of these arguments has been accepted for publication in peer-reviewed scientific journals, and both arguments have been rejected by the scientific community as artificial science.

Evolutionary scientists assert that the main procedure for evolutionary changes in species happens through three biological processes called natural selection, mutation, and genetic drift.

Natural selection is a process by which favorable heritable traits—those that make it more likely for an organism to survive and successfully reproduce—become more common in a population over successive generations. The principles of natural selection are based on three supposedly realistic observations. First, every individual is supplied with hereditary material in the form of genes and DNA, which are passed down from their parents and then are passed on to their offspring. Second, organisms tend to produce more offspring than the environment can support, causing the strong to survive and the weak to become extinct. Third, there are variations among offspring because of either the random introduction of new genes via mutations or the reshuffling of existing genes during sexual reproduction. Creation scientists generally do not support natural selection, because it contradicts the biblical Creation narrative in Genesis.

Mutations are changes in the DNA sequence of a cell. These changes happen in a variety of ways. Sometimes—however, extremely seldom—simple copying errors occur when DNA replicates itself. Every time a cell divides, its DNA is duplicated, so that each of the resulting cells each has a full set of the exact same DNA. DNA damage may be caused by environmental conditions, such as sunlight, cigarette smoke, poisons, and radiation. Cells possess built-in mechanisms that recognize and repair most of the errors that transpire during DNA replication from environmental damage. Mutations, or copying errors, are generally detrimental to the gene instead of beneficial. When errors occur in DNA cells that produce eggs and sperm, they are called germ line mutations, and these can be passed on from one generation to the next. Germ line mutations cause deceases to run in families and are solely responsible for hereditary deceases.

Genetic drift is believed to be an important evolutionary process leading to changes in populations' allele frequencies over time, triggered by random chance rather than the need for adaption or natural selection. In natural selection, allelic frequency is modified on the basis of the fittest genes surviving to reproduce and the weaker genes dying off. Genetic drift tends to be more common in smaller populations; natural selection, on the other hand, is more frequent in larger populations. Genetic drift may also cause gene variations to disappear completely, thereby reducing genetic variability, in contrast to natural selection, which makes gene variants more common or less common depending on their reproductive success. Changes due to genetic drift are not driven by environmental or adaptive pressures and may be beneficial, neutral, or detrimental to reproductive success. Evolutionism is still considered one of the greatest unifying models of modern-day biology. In the last few decades, studies of DNA sequences of closely related species have provided clues to the mutations that produce organisms with different physical features. DNA sequences also identify contemporary organisms that share common evolutionary ancestry.

Microevolution, thus, is evolution on a small scale where changes can and do occur within a single species—a species that can interbreed. The term "microevolution" recently gained a lot of popularity in the antievolution movement, especially by young-Earth Creationists. Selective breeding is a prime example of microevolution. This, for example, is where a farmer breeds selected cows to cause an increase in milk production or to produce a better quality of beef. The fact that humans are growing taller and are getting smarter as time goes by also provides undisputable proof that microevolution does occur.

Creationists find themselves in constant conflict with evolutionists in regard to some of the very basics of the general theory of evolution. Three of these arguments include the following:

1) Origin of life through a biological process starting from nothing
2) Mutations as a credible source of population diversity
3) All life on Earth originate from one common ancestor

Creationists generally feel that there simply is not enough scientific evidence to support any of these aspects of the theory of evolution; they feel that these aspects are, in principle, results of atheistic philosophies. They believe these theories do not acknowledge the existence of God and consequently strongly deny the involvement of a supernatural intelligent designer. Instead, they accept only the biblical version of the origin of life, in which God created all species exactly the way they are, which characterizes a complete contradiction to the theory of macroevolution.

Sadly, both evolution and creation sciences suffer because of misunderstandings about each other's philosophies and the refusal to consider the opposite parties' point of view. Most of us have been led to believe there are only two choices; we believe the biblical views, or we believe science. In truth, there is no reason we must compromise either the Bible or science.

Conflicts between evolutionism and Creationism happen when supporters of either view criticize the other. Evolutionists argue that Creationism is not a scientific theory because it cannot be tested by the scientific method, basically admitting that there is no scientific explanation for God. Creationists argue that for this reason, evolutionists do not consider God or intelligent design in their theories, and that therefore, evolution is nothing more than just another scientific fairy tale, rather than a proven fact. Scientific methods, which are based on physical evidence, can never be reconciled with the Creationist faith-based belief that the Genesis of the Bible is the only true account of the origin of life. Science requires that hypotheses or theories should be testable and supported by physical and observational evidence, whereas religion requires acceptance of a doctrine or belief without scientific proof. For this reason, conflicts between evolution and Creationism can likely never be resolved and will continue forever. There is also the question of exactly how much of their various components are theories only, or actual proven facts.

Homology is a theory that macroevolution can be demonstrated by examining the similarities in the makeup and physiologies of different organisms—for example, comparing the human arm to the wing bones of a bat. A precise definition of homology is simply "Having a common evolutionary origin." Creationists make it quite clear that the homology argument is by no means justifiable. A prime example of a homology argument is that DNA similarities between humans and other living organisms are sufficient evidence to prove the theory of evolution. To be more precise, DNA testing has shown that humans and chimpanzees are 96 percent genetically similar; the evolutionist presents this as overwhelming evidence that apes and humans have a common ancestry, and that man is a specific type of ape. On the other hand, a mouse is also genetically very similar to a human. Creationists dispute science's interpretation of genetic evidence in the study of human evolution. They argue that it is a dubious assumption that genetic similarities between various animals imply a common ancestral relationship, and that scientists are arriving at this interpretation only because they have preconceived notions that such shared relationships must exist. Creationists, rather, argue that genetic mutations are strong evidence against evolutionary theory because genetic mutations are copying mistakes that would almost certainly be detrimental instead of beneficial to the gene.

Young-Earth' Creationists' opinions differ somewhat from Creationists'. Young-Earth Creationists generally believe that speciation does happen; however, they believe it occurs at a much faster rate than evolutionists believe is the case. They do believe that some organisms do evolve over time, and in fact, they do agree to a certain extent with the basic mechanisms of biological evolution as proposed by Charles Darwin. Young-Earth Creationists also agree that the processes of genetic recombination and natural selection can also result in the formation of new species. In fact, they believe that extremely rapid evolution occurred after Noah's Flood to create the species that we see today from the small number of species that were on the ark.

Creationists, as well as intelligent design advocates, insist that the genetic code, genetic programs, and biological information do require an intelligent designer in regard to the origins of life, and they admit this is one of the main conflicts with the theory of evolution. Dr. Walt Brown (b. 1937) is considered to be one of the most notorious leaders of the Creationism movement. He states that the genetic materials that control the biological processes of life are coded information and that human experience tells us that codes are created only as a result of intelligence and not merely by a process of nature. Dr. Brown also asserts that the "information stored in the genetic material of all life is a complex program. Therefore, it appears that an unfathomable intelligence created these genetic programs."

Stephen Jay Gould, a self-proclaimed agnostic, and atheist Richard Dawkins are this decade's most notorious and outspoken proponents of evolution, continually attacking and criticizing Creationism. In his fascinating book *The Language of God*, Francis Collins (b. 1950), one of the most prominent scientists in the world today and also a believer in God, expresses his views on Creationism and evolution.

Stephen Jay Gould (1941–2002) was a famous American paleontologist, evolutionary biologist, and historian of science. He was also one of the most influential and widely read writers of popular science in his generation. Gould spent most of his career teaching at Harvard University and working at the American Museum of Natural History in New York. In the latter years of his life, Gould also taught biology and evolution at New York University near his home in SoHo, New York. He strongly campaigned against Creationism and proposed that science and religion should be considered two distinct fields, or "magisteria," whose authorities do not overlap. Raised in a modest Jewish home, Gould was not very religious and preferred to be called an agnostic. As a vivid supporter of evolutionary theory, Gould wrote many books on the subject, trying to explain his understanding of everyday evolutionary biology to a wide audience. A recurring theme in his writings is the history and development of evolution.

Richard Dawkins (b. 1941) is a British ethnologist, evolutionary biologist, popular science author, and a former chair in the Public Understanding of Science at Oxford University. He became an instant celebrity with the publication of his 1976 book *The Selfish Gene*, which popularized the gene-centered view of evolution. Dawkins is well known for his blunt criticism of Creationism and intelligent design and makes regular television and radio appearances, predominantly discussing and ridiculing these topics. Dawkins is an atheist, a secular humanist, a skeptic, and a rationalist. In the media, he has often been referred to as Darwin's Rottweiler—compared to English biologist T. H. Huxley, who was known as Darwin's Bulldog for his promotion of Charles Darwin's evolutionary ideas. In his 1986 book *The Blind Watchmaker*, Dawkins again profoundly criticized intelligent design, an important argument for Creationism. Dawkins argued against the watchmaker analogy made famous by the eighteenth-century English theologian William Paley in his book *Natural Theology*. Paley argued that just as a watch is too complicated and too functional to have sprung into existence merely by accident, so too must all living things, with far greater complexity, be purposefully designed. According to Dawkins, however, natural selection is sufficient to explain the apparent functionality and nonrandom complexity of the biological world, and it can be said to play the role of watchmaker in nature, notwithstanding an automatic, unintelligent, blind watchmaker. Dawkins is an outspoken atheist and a prominent critic of religion. He has been described as a vocal, militant rationalist and as a militant atheist. Dawkins explains that his own atheism is the logical extension of his understanding of evolution and that religion is not compatible with science. In his 2006 book *The God Delusion*, Dawkins contended that a supernatural creator almost certainly does not exist and that faith qualifies as a delusion—a fixed false belief. He also described the teaching of Creationism to minors as a form of child abuse. How appropriate are the words of king David when he wrote: "The fool hath said in his heart, there is no God. They are corrupt, they have done abominable works; There is none that doeth good" (Psalms 14:1).

Francis Collins (b. 1950), MD, PhD, is an American physician and geneticist noted for his significant discoveries of disease genes and his leadership of the Human Genome Project (HGP). He is described by the Endocrine Society as "one of the most accomplished scientists of our time." He currently serves as director of the National Institutes of Health in Bethesda, Maryland. Collins has written a fascinating book about his Christian faith, and founded and was president of the BioLogos Foundation before accepting the nomination to lead the NIH. On October 14, 2009, Pope Benedict XVI appointed Mr. Collins to the Pontifical Academy of Sciences.

In his fascinating 2006 book *The Language of God*, Collins considers scientific discoveries an "opportunity to worship" and examines and subsequently rejects Creationism and intelligent design. His own belief system is theistic evolution or evolutionary creation, which he prefers to term BioLogos.

During a debate with Richard Dawkins, Dr. Collins stated that God is the explanation of those features of the universe that science finds difficult to explain (such as the values of certain physical constants favoring life), and that God himself does not need an explanation since he is beyond the universe. Dawkins called this "the mother and father of all cop-outs" and "an incredible evasion of the responsibility to explain," to which Mr. Collins responded, "I do object to the assumption that anything that might be outside of nature is ruled out of the conversation. That's an impoverished view of the kinds of questions we humans can ask, such as 'Why am I here?', 'What happens after we die?' If you refuse to acknowledge their appropriateness, you end up with a zero probability of God after examining the natural world because it doesn't convince you on a proof basis. But if your mind is open about whether God might exist, you can point to aspects of the universe that are consistent with that conclusion." However, as a God-believing scientist, Mr. Collins does remain firm in his rejection of intelligent design.

During the last century, there have been many significant negative social ramifications of the adoption of the theory of evolution. Evolutionary theories laid the foundations for Social Darwinism, Nazism, Communism, and racism.

Devoted evolutionist Stephen Jay Gould admitted the following about Ernest Haeckel (1834–1919), a controversial German biologist and artist: "Haeckel was the chief apostle of evolution in Germany ... His evolutionary racism, his call to the German people for racial purity and unflinching devotion to a 'just' state. His belief that harsh, inexorable laws of evolution ruled human civilization and nature alike, conferring upon favored races the right to dominate others, the irrational mysticism that had always stood in strange communion with his brave words about objective science—all contributed to the rise of Nazism."

Adolf Hitler (1889–1945), leader of the German Nazi Party and architect of the Second World War, wrote the following evolutionary racist material in his book *Mein Kampf*: "If nature does not wish that weaker individuals should mate with the stronger, she wishes even less that a superior race should intermingle with an inferior one. Because in such cases, all her efforts, throughout hundreds of thousands of years, to establish an evolutionary higher stage of being, may thus be rendered futile." Hitler also wrote, "The stronger must dominate and not blend with the weaker, thus sacrificing his own greatness. Only the born weakling can view this as cruel, but he, after all, is only a weak and limited man; for if this law did not prevail, any conceivable higher development (Hoherentwicklung) of organic living beings would be unthinkable."

Noted evolutionary Scottish anthropologist Sir Arthur Keith (1866–1955) conceded the following in regard to Hitler and his theory of evolution: "The German Fuhrer, as I have consistently maintained, is an evolutionist; he has consciously sought to make the practices of Germany conform to the theory of evolution." Richard Dawkins stated the following regarding Adolf Hitler in an interview: "What's to prevent us from saying Hitler wasn't right? I mean, that is a genuinely difficult question." The interviewer commented, "I was stupefied. He had readily conceded that his own philosophical position did not offer a rational basis for moral judgments. His intellectual honesty was refreshing, if somewhat disturbing on this point."

A. E. Wilder-Smith (1915–1995), a Young-Earth Creationist and chemist, wrote the following regarding Nazism and the theory of evolution: "One of the central planks in Nazi theory and doctrine was evolutionary theory ... [that] all biology had evolved upward, and that less evolved types should be actively eradicated and that natural selection could and should be actively aided. Therefore, the Nazis instituted political measures to eradicate Jews, and blacks, whom they considered as 'underdeveloped.'

Dr. Josef Mengele (1911–1979) was a Nazi SS officer and physician in the concentration camp Auschwitz-Birkenau. His evolutionary thinking was parallel with social Darwinist theories that Adolph Hitler and a number of German academics found appealing. Mengele, referred to as the Angel of Death, studied under the leading proponents of the "unworthy life" branch of evolutionary thought. Mengele was one of the most notorious criminal individuals associated with Nazi death camps and the Holocaust. Mengele obtained his infamous reputation because of his crude experiments on twins while a physician at the Auschwitz-Birkenau death camp.

Adolf Eichmann (1906–1962) was a high-ranking Nazi and SS *Obersturmbannführer* (lieutenant colonel), and a proponent of evolutionary theory. Because of his organizational skills and ideological reliability, he was tasked to organize and manage the mass deportation of Jews to ghettos and extermination camps in Nazi-occupied Eastern Europe, and he worked under Ernst Kaltenbrunner until the end of the war. After the war he fled to Argentina, where he was captured by Israeli Mossad agents and indicted by Israeli court on fifteen criminal charges, including charges of crimes against humanity and war crimes. He was convicted and hanged.

Hannah Arendt (1906-1975) was one of the most influential philosophers of the twenties century. Born into a German-Jewish family, she was forced to leave Germany in 1933 and lived in Paris for the next eight years, working for a number of Jewish refugee organizations. In 1941 she immigrated to the United States and soon became part of a lively intellectual circle in New York. In 1963, she made the following comment about Eichmann, "Adolf Eichmann, the sinister figure who had charge of the extermination program, has estimated that the anti-Jewish activities resulted in the killing of 6 million Jews. Of these, 4 million were killed in extermination institutions, and 2 million were killed by Einsatzgruppen, mobile units of the Security Police and SD which pursued Jews in the ghettos and in their homes and slaughtered them by gas wagons, by mass shooting in antitank ditches, and by every device which Nazi ingenuity could conceive. So thorough and uncompromising was this program that the Jews of Europe as a race would no longer exist, fulfilling the diabolic 'prophecy' of Adolf Hitler at the beginning of the war." (Arendt 1963)

Ironically, Hitler died in much the same way as most of his Jewish victims; he was shot (by himself, committing suicide), doused with gasoline, and burned. About four thousand years ago God spoke the following words to Abraham: "And I will bless them that bless thee, and him that curseth thee will I curse: and in thee shall all the families of the earth be blessed" (Genesis 12:3).

The controversy between Creationist and evolutionist has been raging ever since Darwin's publication of his *Origin of Species* in 1859, and as long as there are atheist scientist and God-believing Christians, the controversy will continue; there is no end in sight. Tragically, some of these controversies, as bizarre as it may seem, made it all the way to the Supreme Court of the United States. Should schools be allowed to teach scientific Creationism? Should schools be allowed to teach about evolution? Is evolution a religion? These are some of the many questions courts at all levels have had to deal with as a result of the attempts by one party to suppress the other's doctrines. Creationists try to stop the teaching of evolution and replace it with the teaching of their own religious beliefs, and evolutionists try to forbid the teaching of Creation. If evolution is taught in public school science classes, shouldn't other theories about the origins and development of life also be taught at the same time? Isn't the focus on just one idea narrow-minded? Most believe that it is and therefore argue that there should be at least a balance; if the theory of evolution is taught, then Creationism should also be taught. It is indeed hard to imagine that it takes a supreme court of law to decide if either evolution or Creationism is permitted to be taught in public schools. Is it not true that education is a quest for knowledge? Is it not true that education is a quest for the truth? For example, both Jesus Christ and Charles Darwin were historical figures; they both made a tremendous impact on humanity. Why should the truth not be taught about both instead of one, without suppressing the truth about the other?

In North America we are very fortunate indeed; we still may exercise our freedom of religion. We can choose any religion we desire, no matter if it be Protestant, Catholic, Baptist, Muslim, or Buddhist. Nevertheless, we must realize that if we want to exercise our right to religious freedom, we must also let the other party have *their* right to *their* religion!

Here is a summary of some of the most absurd and notorious court cases regarding Creation vs. evolution:

The Scopes Trial was an American legal case in 1925 in which high school biology teacher John Scopes was accused of violating the state's Butler Act. The Butler Act made it "unlawful for any teacher or other instructor in any university, college, normal, public school or other institution of the state which is supported in whole or in part from public funds derived by state or local taxation to teach the theory or doctrine that mankind ascended or descended from a lower order of animals. And also, that it be unlawful for any teacher, textbook commission, or other authority exercising the power to select textbooks for above-mentioned institutions to adopt or use in any such institution a textbook that teaches the doctrine or theory that mankind ascended or descended from a lower order of animal."

Scopes was charged May 5, 1925, with teaching evolution from a chapter in a textbook that showed ideas developed from Charles Darwin's book *On the Origin of Species*. The Butler Act actually did not violate any church or state laws; it merely prohibited the teaching of evolution on the grounds of intellectual disagreement, leaving Creationism as the only option. Scopes was found guilty, but the verdict was overturned on a technicality and he was never punished. The trial drew intense worldwide publicity, as national reporters flocked to the small town of Dayton, Tennessee, to cover and sensationalize the big-name lawyers representing each side. William Jennings Bryan, a devoted Baptist and three-time presidential candidate for the Democratic Party argued for the prosecution, while famous defense attorney Clarence Darrow, an agnostic, spoke for Scopes. The trial positioned Modernists, who believed religion was consistent with evolution, against Creationists, who contended that the Word of God as revealed in the Bible undermined all human knowledge. The trial was thus both a religious and theological contest, and it was aimed to set precedence for the creation-evolution controversy. After Scopes was convicted, Creationists throughout the United States wanted similar antievolution laws for their states. However, the teaching of scientific evolution continued to expand, while Creationists kept trying to utilize state laws to reverse the trend. By 1927 there were thirteen states, both in the North and the South that had implemented some form of antievolution law. At least forty-one bills of resolution had been introduced into state legislatures, with some states facing the same issues repeatedly; however, almost all of these efforts were rejected.

The case of Epperson v. Arkansas focused on the constitutionality of a 1928 Arkansas statute prohibiting the teaching of human evolutionary theory in its public schools and universities. The Arkansas statute was modeled after Tennessee's 1925 Butler Act, the subject of the well-known Scopes Trial in 1925. The Tennessee Supreme Court upheld the constitutionality of the Tennessee law, allowing the state to continue to prohibit the teaching of evolution.

This famous case involved the teaching of biology in a Little Rock high school about forty years after the Scopes Trial. Based on the recommendation of the school's biology teachers, the administrators adopted a new textbook for the 1965–1966 school year that contained a chapter discussing Charles Darwin and his evolutionary theory, and prescribed the subject to be taught to the students.

Susan Epperson was employed to teach tenth-grade biology at the Little Rock Central High School. The adoption of the new textbook and curriculum standard put her in a legal dilemma because it remained a criminal offense to teach evolutionary material in her state. To comply with instructions from her employer, the school district would put her at risk of dismissal. Epperson was not at all opposed to the teaching, and with backing from the Arkansas chapter of the National Education Association and the American Civil Liberties Union, and the indisputable support of the Little Rock Ministerial Association, she filed suit to test the federal constitutionality of the Arkansas state law. She filed in the chancery court in Pulaski County, seeking nullification of the law and an injunction against her being dismissed for teaching the evolutionary curriculum.

The chancery court decided that the statute violated the Fourteenth Amendment to the US Constitution, which protects citizens from state interference with freedom of speech and thought as contained in the constitution's First Amendment. This lower court decided the law was unconstitutional because it "tends to hinder the quest for knowledge, restrict the freedom to learn, and restrain the freedom to teach," meaning Susan was free to teach evolutionary theory to her class without the chance of being reprimanded. However, in 1967, the Arkansas Supreme Court reversed the lower court ruling. The opinion was just two sentences long and offered little explanation for its reversal. This decision left the ban against teaching evolution in effect, meaning evolutionary teaching was not legal in Arkansas.

Susan appealed the State Supreme Court's reversal to the US Supreme Court. Eugene R. Warren presented arguments for the accuser, Epperson; and Don Langston, an assistant attorney general for Arkansas, argued on behalf of the state. Both Langston and the state appeal court focused on the power given to states to set curriculum standards and avoided discussing the evolutionary theory itself; nor did they discuss the boundaries between church and state. The US Supreme Court found the reasons given in the Arkansas reversal were in error. They went on to say that the clear purpose of the Arkansas statute against the teaching of evolution was to protect a particular religious view, and was thus unconstitutional. The court found that not only was the state prohibited from advancing or protecting a particular religious view, but also that the state had no legitimate interest in protecting any or all religions from views distasteful to them. Justice Hugo Black issued a separate opinion to overturn the Arkansas law, finding that the law was unconstitutionally "vague" rather than an unconstitutional religious infringement. While he agreed with the majority to reverse the state appeal court decision, his opinion details his dissent from the majority over the First Amendment issue.

The standard set in the Epperson trial, in which the court concluded the sole motive behind the ban against evolution teaching in Arkansas was to protect a particular religious view, effectively nullified all other related evolution education prohibitions throughout the United States. Within a short time of the Epperson decision, religious opponents of the teaching attempted through other means to lessen its influence in the curriculum, including requiring schools to teach biblical creation alongside evolution or forcing schools to provide disclaimers stating that evolution was "only a theory."

Daniel v. Waters was a 1975 legal case in which the US Court of Appeals for the Sixth Circuit struck down Tennessee's law regarding the "equal time" teaching of evolution and Creationism in public-school science classes because it violated the Establishment clause of the US Constitution. Following this ruling, Creationism was stripped of explicit biblical references and renamed "creation science," and several states passed legislation requiring that creation science be given equal time with the teaching of evolution.

Kitzmiller v. Dover Area School District was the first direct challenge brought in the US federal courts against a public school district that required the teaching of creation science, renamed "intelligent design" for political reasons, as an alternative to evolution. The plaintiffs successfully argued that intelligent design is a form of Creationism and that the school board policy thus violated the Establishment Clause of the First Amendment to the US Constitution. On December 20, 2005, Judge Jones issued his 139-page findings of fact and decision, ruling that the Dover mandate was unconstitutional and barring intelligent design from being taught in Pennsylvania's Middle District public-school science classrooms. The eight Dover school board members who had voted for the intelligent design requirement were all defeated in a November 8, 2005, election by challengers who opposed the teaching of intelligent design in a science class, and the current school board president stated that the board did not consider appealing the ruling.

Judge Jones himself anticipated that his ruling would be criticized, saying in his decision that "those who disagree with our holding will likely mark it as the product of an activist judge. If so, they will have erred as this is manifestly not an activist Court. Rather, this case came to us as the result of the activism of an ill-informed faction on a school board, aided by a national public interest law firm eager to find a constitutional test case on ID, who in combination drove the Board to adopt an imprudent and ultimately unconstitutional policy. The breathtaking insanity of the Board's decision is evident when considered against the factual backdrop, which has now been fully revealed through this trial. The students, parents, and teachers of the Dover Area School District deserved better than to be dragged into this legal maelstrom, with its resulting utter waste of monetary and personal resources."

Darwin's theory of evolution is a theory in crisis in light of the tremendous advances made in molecular biology, biochemistry, and genetics over the past fifty years. Every time new scientific discoveries are made, they always seem to discredit evolution, thus advantageous to Creationism. In his 1985 book *Evolution: A Theory in Crisis*, molecular biologist Michael Denton questioned the validity of Neo-Darwinism and argued that evidence of divine design does exist in nature. The book was instrumental in starting the intelligent design movement. Denton was an influential proponent of intelligent design and was a former senior fellow of the Discovery Institute's Center for Science and Culture, hub of the intelligent design movement. Over the last decade, science has discovered that there are in fact tens of thousands of irreducible complex systems on the cellular level. Denton (b. 1943), a British Australian science author and biochemist, writes, "Although the tiniest bacterial cells are incredibly small, weighing less than 0.10^{12} grams, each is in effect a veritable micro-miniaturized factory containing thousands of exquisitely designed pieces of intricate molecular machinery, made up altogether of one hundred thousand million atoms. Far more complicated than any machinery built by man and absolutely without parallel in the non-living world ... And we don't need a microscope to observe irreducible complexity." Creation Ministries International describes irrational evolutionary thinking as follows: "Underpinning this abandonment of faith in God is the widespread acceptance of evolutionary thinking—that everything made itself by natural processes; that God is not necessary. There is obvious 'design', they admit, but no designer is necessary. The 'designed' thing designed itself! This philosophy, where the obvious evidence for God's existence is explained away, naturally leads to atheism and secular humanism (man can chart his own course without God). Such thinking thrives in universities and governments institutions today."

Fossils

When Darwin published his *Origin of Species* more than 150 years ago, geologists had only recently discovered fossils. The existence of fossilized plants, animals, and even exotic creatures like dinosaurs challenged many scientific and religious theories. Darwin's supporters looked to the fossil record for support for his theory of evolution; however, there was none to be found. Darwin himself admitted that it was essential that fossils of ancestral creatures must be found. He also confessed that so far none had been found, but he believed eventually that they would be; otherwise, his theory could be disproved and be rendered worthless. The desperate need for a fossilized "missing link" caused thousands of geologists and scientists, professionals and amateurs alike, to join the cause and search the globe for fossilized remains, causing the era of the bone hunters, who searched for fame and fortune.

There are well over a hundred million fossils in museums and private collections in the world today—about forty million alone in the Smithsonian Natural History Museum. If Darwin's theory were actually correct, there would be tens of millions of unquestionable transitional fossils available in support of evolution. However, not even one fossil that can provide evidence that one organism evolved into another has ever been produced. For example, a consecutive series of fossils that suggest that the fins of a fish evolved into the legs of a mammal simply do not exist. Darwin himself proposed that the fossil record could be used to test his evolutionary theory; however, the available evidence in the fossil records conclusively flunks his test even today. The fossil record is often employed as evidence in the creation-versus-evolution controversy. However, the record does not show support for evolution; it represents one of the major flaws of the theory of evolution. Despite the large number of fossils available to scientists in 1981, evolutionist Mark Ridley, who currently serves as a professor of zoology at Oxford, commented that in any case, no real evolutionist, whether gradualist or punctuationist, uses the fossil record as evidence in favor of the theory of evolution as opposed to special creation ... Fossil evidence of human evolutionary history is fragmentary and open to various interpretations. Fossil evidence of chimpanzee evolution is absent altogether.

Darwin described his desperation for the discovery of these fossils in his *Origin of Species* with the following statements: "Firstly, why, if species have descended from other species by insensibly fine gradations, do we not everywhere see innumerable transitional forms? Why is not all nature in confusion instead of the species being, as we see them, well defined... But, as by this theory, innumerable transitional forms must have existed, why do we not find them embedded in countless numbers in the crust of the earth?... Lastly, looking not to any one time, but to all time, if my theory be true, numberless intermediate varieties, linking closely together all the species of the same group, must assuredly have existed ... Why then is not every geological formation and every stratum full of such intermediate links? Geology assuredly does not reveal any such finely graduated organic chain; and this, perhaps is the most obvious and gravest objection which can be urged against my theory."

Stephen Jay Gould, one of the most outspoken proponents of evolution confessed that "the extreme rarity of transitional forms in the fossil record persists as the trade secret of paleontology." The apparent nonexistence of transitional fossils provides a severe blow to the evolutionary theory, providing massive evidence to favor Creationism. The theory of evolution is based on the "survival of the fittest"; evidently, it is also the theory that has the most difficulty surviving!

The following appeared in *Newsweek* magazine: "The missing link between man and apes, whose absence has comforted religious fundamentalists since the days of Darwin, is merely the most glamorous of a whole hierarchy of phantom creatures ... The more scientists have searched for the transitional forms that lie between species, the more they have been frustrated" ("Is Man a Subtle Accident," *Newsweek*, November 3, 1980).

Obviously, evolution is not capable of producing any sort of link that could indicate how everything could evolve from nothing, how life could evolve from non-life, or how one kind of creature could evolve into a completely different kind, lacking the genetic coding to do so. Genesis of the Bible simply states that all species were created to "their own kind."

News reports regularly feature evolutionary missing link stories as supposed evidence for either human or animal evolution. Lucy, Nebraska Man, Java Man, Piltdown Man, and Neanderthal Man have all been identified as missing links at one time or another. Nebraska Man turned out to be nothing more than the tooth of some sort of swine, Piltdown Man turned out to be the most notorious evolution-related frauds in history, and all others turned out to be either the remains of human beings or apes. No evidence for a human missing link has ever been found.

The coelacanth is a strange-looking sort of fish that many paleontologists considered, based on the fossil record, to have gone extinct some two hundred million years ago. Darwin theorized that amphibians like salamanders, frogs, toads, and crocodiles originally evolved from this bizarre-looking fish. The Coelacanth was considered one of the most important transitional fossils—a fish swimming in the sea that evolved into a creeping mammal. On an expedition to Africa, a French scientist happened to spot a specimen at a local fish mart in Zanzibar, off the coast of East Africa. He purchased the fish, took it home, conducted further studies and found that the coelacanth was not extinct at all and actually was considered quite a delicacy among the natives. Instead of a two-hundred-million-year extinct species, it turned out to be nothing more than just another fish, alive and well. He did comment that it was quite an ugly thing, but nevertheless, it tasted pretty good!

Lucy is the popular name given to a famous fossil skeleton that American anthropologist Donald Johanson found in Ethiopia in 1974.

To many people, Lucy is still regarded as some kind of missing link between apelike creatures and humans, thus supposedly providing evidence for evolution. According to Richard Leakey, who, along with Donald Johanson, became probably the best-known fossil anthropologists in the world, Lucy's skull was so incomplete that most of it was "imagination, made of plaster of Paris." Leakey admitted in 1983 that no firm conclusion could be drawn about what species Lucy belonged to. Neither Lucy nor any other australopithecine should therefore be considered an intermediate between humans and African apes. They are not similar enough to us humans to be considered any sort of ancestor of ours. Lucy and the australopithecines reveal absolutely nothing about human evolution and should not be considered to have any sort of missing-link status. Three scientists from the departments of anatomy, anthropology, and zoology at Tel Aviv University in Israel reported that the jawbone of Lucy is a close match to that of a gorilla. To make things even more complicated, Richard Leakey has recently discovered a humanoid skull even older than the ones that made him famous, but much more modern in type. He commented that "either we toss out the skull, or we toss out our theories of early man." The Creationist alternative—that humans, apes, and all other creatures were created exactly the way they are right from the beginning—remains a plausible explanation consistent with all the evidence.

Nebraska Man was named in 1922 from a humanlike tooth found in Nebraska in the Pliocene deposits. Harold Cook, a rancher and geologist, found the tooth in 1917, and in 1922 he sent it to Henry Fairfield Osborn (1857–1935), a paleontologist and also the president of the American Museum of Natural History, for further analysis. Osborn originally identified the tooth as an ape's and quickly published a paper identifying it as a new species, which he named *Hesperopithecus haroldcookii*. Most scientists were skeptical even of the modest claim that the *Hesperopithecus* tooth did belong to a primate. A popular British magazine, the *Illustrated London Times*, commissioned an illustrator, collaborating with scientist Grafton Elliot Smith, to produce precise drawings of Nebraska Man and his family. Nebraska man was obviously more sensationalized in Europe than in North America, and was propagandized as a missing link—an early human ancestor. However, further excavations revealed that the tooth actually belonged to a peccary, an animal similar to (and closely related to) a pig.

As Creationists interpret the story, evolutionists used this one tooth to build an entire species of primitive man, complete with visionary illustrations of Nebraska Man and his family, and often claim that Nebraska Man was used as proof of evolution during the Scopes Monkey Trial in 1925, However, this claim is not quite true. There was no such scientific evidence presented at this trial. Although some evidence was read into the trial record, it was done so without reference to Nebraska Man. It is also not quite true that Nebraska Man was widely accepted as an ape-man, or even as an ape, and scientific response of the time was insignificant. Few, if any, scientists actually declared Nebraska Man to be a human ancestor. The scientific community, including Osborn and his colleagues, later identified it only as an advanced primate of some sort. Osborn, in fact, specifically avoided making any extravagant claims about *Hesperopithecus* being an ape-man or human ancestor. He said, "I have not stated that Hesperopithecus was either an Ape-man or in the direct line of human ancestry, because I consider it quite possible that we may discover anthropoid apes with teeth closely imitating those of man … Until we secure more of the dentition, or parts of the skull or of the skeleton, we cannot be certain whether Hesperopithecus is a member of the Simiidae (Apes) or of the Hominidae. (Humans)"

Although Nebraska Man is often ridiculed by Creationists, it should not be considered an embarrassment to science. The scientists involved may have made hastily decisions, were mistaken, and were perhaps somewhat sloppy, but they were not dishonest. The Nebraska Man episode is actually an excellent example of how the scientific process is supposed to function. When the scientists discovered a problem with proper identification, they conducted further investigation, found data that contradicted their earlier opinions, and then promptly abandoned those opinions.

A similar discovery, which was also based upon a tooth, was Southwest Colorado Man. It also became evident that this particular tooth actually did not belong to a swine; this time it belonged to a horse!

One of the most famous missing links ever discovered was the anthropoid Java Ape-Man, named *Pithecanthropus erectus* (meaning "erect ape-man"). Java Man was discovered on the banks of the Solo River in 1891 in east Java, Indonesia, by a Dutch scientist, Dr. Eugene Dubois (1858–1940), an enthusiastic evolutionist and former student of Ernest Haeckel. Dr. Dubois's find consisted of a small piece of the top of a skull, a fragment of a left thighbone, and three molar teeth. (A leading authority later identified two of the teeth as those of an orangutan and the other as human.) Although the evidence was considered substantial, it was still fragmentary. Furthermore, these remnants were not found in one general location; they were collected over a range of about forty-five (some say sixty-five) feet. Also, they were not discovered simultaneously, but over a period of about one year. To make things more complicated, these remains were found in an old riverbed, mixed with the bones of other extinct animals. Despite all of these controversies, evolutionists continue to insist that Java Ape-Man lived about 750,000 years ago. For a full thirty years, Dubois conveniently neglected to announce to the scientific community that he had also found two human skulls in the same general location.

The "scientific experts" insisted that from these minor fragments, an entire prehistoric race could be reconstructed. However, this certainly raises more questions than it provides answers. For instance, how is it possible to reconstruct such a complete detailed specimen, with such confidence, from such unsubstantiated evidence? How can the "experts" be so certain that all this evidence originated from the same species? How could these unpetrified bones have managed to survive for 750,000 years without disintegrating? Well, as it turns out, even the opinions of the "experts" about the identification of these fossil fragments differed greatly. Twenty-four of the most notorious European scientists organized a meeting to examine and evaluate the extraordinary find; ten of them concluded they came from an ape, seven insisted they came from a man, and seven said they did not belong to a missing link. Controversy and division surrounded the discovery. Renowned professor Virchow of Berlin admitted, "There is no evidence at all that these bones were parts of the same creature."

Even Dr. Dubois himself later recanted his own opinion. His final conclusion was that the bones were the remains of a gibbon, a type of ape that makes its habitat in tropical and subtropical rainforests, including the Indonesian islands of Sumatra, Borneo, and Java.

Another *Pithecanthropus* was found in Java in 1926. This discovery was also billed as an extraordinary breakthrough, the missing link for sure. It was later exposed to be the knee bone of an extinct elephant.

The story of the Piltdown hoax is considered the most notorious in the desperate search for the missing link. For decades, Piltdown Man had been proclaimed the ultimate, undisputable proof that man evolved from apes.

The remains of Piltdown Man were allegedly discovered sometime before 1912 by Charles Dawson, a Sussex lawyer and amateur paleontologist. He produced some bones, teeth, and primitive implements, which he said he had discovered in a gravel pit at Piltdown, Sussex, England. It is believed that none other than Sir Arthur Conan Doyle, the imaginative author of the Sherlock Holmes detective mystery stories, was conspiring with Dawson, originating the idea for this fraudulent placement and later "discovery" of the bones. Dawson took the bones to Dr. Author Smith Woodword, an eminent paleontologist at the British Museum. This "magnificent discovery" came at a most opportune time. Both Charles Darwin and Darwin's "Bulldog," Thomas Huxley, had both passed away, and although "fossilized human bones" had been found in other places, such as the Neanderthal in Germany, none of them really contributed to the cause of evolution. They were all clearly human remains. Woodward was an avid paleontologist, and had written many papers on fossilized fish. Dawson and Woodward had many lengthy discussions over the bones and spent considerable time studying them. Then Arthur Keith, an anatomist, was invited to join the team. Keith was one of the most highly respected scientists in England at that time. The author of several classic works, he has all the credentials of respectability; he holds a doctorate in medicine, is a fellow of the Royal College of Surgeons, is president of the Royal Anthropological Institute, and holds memberships in the Anatomical Society and the British Association for the Advancement of Science. Grafton Elliot

Smith, a renowned brain specialist, also joined the team. So here was a team of scientists gathered together that was the most respected in the British Isles as well as around the world. A flood of doctoral studies were performed on Piltdown Man, and the remains were confirmed by this team of exquisite scientists to be about three hundred thousand years old. Indisputably, this find was considered to stand the test of time and to establish evolution as an unquestionable fact of science. Here at last, they triumphantly declared, was the long-awaited missing link. A whole generation grew up with Piltdown Man as their purported ancestor. Textbooks, exhibits, displays, encyclopedias—all spread the news that we as humans evolved from apes after all.

Decades passed, and then the whole story blew apart. Suddenly, in October of 1956, the entire hoax was exposed. Reader's Digest published an article, summarized from *Popular Science Monthly*, entitled "The Great Piltdown Hoax." Sir Solly Zuckerman, an expert in the field, commented that the person or persons who perpetrated this deliberate and unscrupulous hoax knew more about ape bones than the scientists at the British Museum.

Fluorine testing is a method of determining whether different bones were buried at the same time or at different times. This is performed by measuring the amount of fluorine the bones have absorbed from groundwater. Employing this new method, the Piltdown bones were determined to be fraudulent. Further critical investigation revealed that the jawbone actually belonged to an ape that had died only fifty years earlier. Some of the teeth had been filed down, and both the teeth and bones had been discolored with dichromate of potash to conceal their correct identity. Piltdown Man was built upon a deception that completely fooled all the "experts" who promoted Piltdown Man with the greatest confidence.

Pierre Teilhard de Chardin, a paleontologist and Jesuit theologian, was also involved in the hoax. Teilhard authored several philosophical books in which he attempted to harmonize evolution and Christianity. Frustrated by the lack of convincing evidence for Darwin's theory of evolution, Teilhard was apparently motivated into promoting the theory by assisting Dawson to fabricate the much-needed missing link. And so, once again, the sincerity of "expert testimony" is called into question.

The first Neanderthal fossils were found in August 1856, three years before Charles Darwin's *Origin of Species* was published. The fossils were found in a limestone quarry near Düsseldorf in the Neanderthal, West Germany. The specimen, dubbed Neanderthal 1, consisted of a skull cap, three right arm bones, two left arm bones, part of the left ilium (the bone forming the upper part of each half of the pelvis), and fragments of a shoulder blade and some ribs. The bones were discovered by workers who originally thought the remains might have belonged to a bear, and they handed them over to amateur naturalist Johann Karl Fuhlrott. Fuhlrott turned the fossils over to anatomist Hermann Schaaffhausen. In 1857, the discovery was jointly announced. Neanderthal Man was described as a semi erect, barrel-chested, rough individual, originally considered an intermediary link between man and apes. With the discovery of other Neanderthal skeletons, it quickly became quite evident that Neanderthal Man was fully erect and also fully human. In fact, his cranial capacity even exceeded that of modern man by more than 13 percent. Misconceptions about Neanderthal skeletons were attributed to two factors: first, the preconception of preprogrammed evolutionary anthropologists who reconstructed him, and second, the fact that the particular individual on whom the initial evaluation was made was crippled with osteoarthritis and rickets.

Today Neanderthal Man is classified as *Homo sapiens*, completely human.

Evolutionary scientists attempt to make us believe that all life on Earth started with a single-celled unit, called a microbe, or that through some sort of spontaneous generation, an early form of life came into existence. From that point on this "thing" evolved into the complexities of life as we know it today, including human beings, without the benefit of any sort of organization, purpose, or intelligent design.

Spontaneous generation is an obsolete theory referring to the process by which life would systematically emerge from sources other than seeds, eggs, or sexual reproduction. Until about 150 years ago, this process was considered an everyday occurrence. Shakespeare discusses snakes and crocodiles forming from the mud of the Nile. In1668, Francesco Redi challenged the idea that maggots arose spontaneously from rotting meat. It was also believed that rats would emerge from dirt thrown into the street. This bizarre theory was finally conclusively disproven in 1859 by the experiments of French biologist Louis Pasteur. Pasteur expanded upon the experiments of other scientists before him. Ultimately, spontaneous generation was succeeded by germ and cell theories.

Chemist Dr. Grebe (1900–1984) commented, "That organic evolution could account for the complex forms of life in the past and the present has long since been abandoned by men who grasp the importance of the DNA genetic code."

In his 1984 book *Darwin was wrong,* Researcher and mathematician I. L. Cohen stated: "At that moment, when the DNA/RNA system became understood, the debate between Evolutionists and Creationists should have come to a screeching halt. ... the implications of the DNA/RNA were obvious and clear. Mathematically speaking, based on probability concepts, there is no possibility that Evolution was the mechanism that created the approximately 6,000,000 species of plants and animals we recognize today."

In his 1985 book *Evolution, a theory in crisis,* Michael Denton stated: "The complexity of the simplest known type of cell is so great that it is impossible to accept that such an object could have been thrown together suddenly by some kind of freakish, vastly improbable, event. Such an occurrence would be indistinguishable from a miracle."

Despite the fact that he was an atheist, scientist Fred Hoyle agreed with Creationists on this point. He repeatedly said that supposing the first cell originated by chance is like believing "a tornado sweeping through a junk-yard might assemble a Boeing 747 from the materials therein ... The notion that ... the operating program of a living cell could be arrived at by chance in a primordial soup here on the Earth is evidently nonsense of a high order." He also compared the probability of life arising by chance to lining up 10^{50} (ten with fifty zeroes after it) blind people, giving each one a scrambled Rubik's Cube, and finding that they all solve the cube at the same moment.

British molecular biologist Francis Crick (1916–2004) was winner of the Nobel Prize in Biology for his work with the DNA molecule in 1962. In 1982 he declared, "An honest man, armed with all the knowledge available to us now, could only state that in some sense, the origin of life appears at the moment to be almost a miracle, so many are the conditions which would have had to have been satisfied to get it going"

Many, if not most, origin-of-life scientists tend to agree with Fred Hoyle: Life could not possibly have originated by accident or some unknown natural process. Some Evolutionists are now probing for some imaginary force within matter that can elevate matter toward the assembly of greater complexity. Creationists believe this theory is doomed to fail, since it contradicts the second law of thermodynamics. Information within DNA molecules is not produced by any known natural interaction of matter. Matter, as well as molecules, has no inherent intelligence; it does not self-organize into codes. Just like a computer, DNA has no intelligence. A computer is only a man-made machine; it only follows instructions, and a computer does not design its own instructions. Complex software programs do not originate within themselves; they require an outside source of intelligence, such as a programmer, to tell the computer what to do and when to do it. Likewise, for DNA, it seems clear that intelligence existed prior to the existence of DNA.

During all recorded human history, there has never been a substantiated case of a living thing being produced from anything other than another living thing. As of yet, evolutionism has not produced a scientifically credible explanation for the origin of such immense complexities as DNA, the human brain, and many other complex elements of the cosmos. It is presumptuous for evolutionists to claim that all living things evolved into existence, when science has yet to discover how even one protein molecule could actually have come into existence by natural processes. There is no scientific proof that life did (or ever could) evolve into existence from nonliving matter. Only DNA is known to produce DNA. No chemical interaction of molecules has even come close to producing this ultra-complex code, which is essential to all known life. Despite these facts, some evolutionists still insist that life evolved from one common ancestor. This means that there must be a relationship between a tulip and a hippopotamus, and an ape must be somehow related to a mosquito, which could have evolved from a banana. The question is, at what point in time this early life form decided to make a split from, let's say, vegetation to a fish or mammal. What was the evolutionary process, or did this also happen strictly by accident alone? At what point did these early life forms, lacking any source of intelligence, decide to develop an eye, to be able to observe the environment around them, and at which point did it decide to develop two eyes, some distance apart, to be able to judge distance and have depth perception? How many thousands, or more likely millions or perhaps billions, of years would it take for the eye to develop through mutations, and most importantly, how did the species see during the process of development? And at which point did species decide on a reproductive system, either through seeds, eggs, or sexual reproduction, and how did they reproduce while their reproductive system was under development? What about the complexity of the brain, the ear, the heart and all other organs? How did these complexities come about? Just accidentally? Not a chance!

This leads us to the age-old question, "Which came first, the chicken or the egg"? The question about the first chicken or egg also evokes not only the question of how life on Earth began but also the question of how the universe originated. An egg is required to hatch an egg-laying chicken, while a chicken is required to lay a chicken-hatching egg. The biblical story of Genesis describes God creating birds and commanding them to multiply, but there is no direct reference to eggs. "And God said, Let the waters swarm with swarms of living creatures, and let birds fly above Earth in the open firmament of heaven And God created the great sea-monsters, and every living creature that moveth, wherewith the waters swarmed, after their kind, and every winged bird after its kind: and God saw that it was good. And God blessed them, saying, be fruitful, and multiply, and fill the waters in the seas, and let birds multiply on the earth" (Genesis 1:20–22). The narrative of Genesis would obviously place the chicken before the egg. In other words, God created an egg-laying chicken rather than a chicken-hatching egg.

It took intelligent mankind close to six thousand years before mankind even discovered the complexities of how the eye actually functions. How could such intricate, complex organ possibly come into being without any plans or the benefit of any source of intelligence? The eye, the ear, and the heart are all examples of irreducible complexities, and there are thousands more, although they were not recognized as such in Darwin's time. Nevertheless, even Charles Darwin himself confessed, "To suppose that the eye with all its inimitable contrivances for adjusting the focus to different distances, for admitting different amounts of light, and for the correction of spherical and chromatic aberration, could have been formed by natural selection seems, I freely confess, absurd in the highest degree."

Some theories brought forth by the scientific community are of such absurd and bizarre nature that they seem to belong on the cover of the *National Enquirer* rather than in a scientific journal. Some scientists even resort to science fiction to propose explanations for the origin of life by suggesting that life started somewhere else in the universe and was then transported and planted on Earth by aliens. "And God said, Let the earth put forth grass, herbs yielding seed, *and* fruit-trees bearing fruit after their kind, wherein is the seed thereof, upon the earth: and it was so. And the earth brought forth grass, herbs yielding seed after their kind, and trees bearing fruit, wherein is the seed thereof, after their kind: and God saw that it was good" (Genesis 1:11–12).

Verses 11 and 12 do not state that God specifically created plant life by His word. They just say, "Let the earth put forth grass, herbs yielding seed, and fruit-trees bearing fruit after their kind, wherein is the seed thereof." "Let the earth put forth" could very well mean that there was some sort of biological system in progress. However, it is quite clear that God caused grass, plants, and trees to be put forth "wherein is the seed thereof," meaning God caused vegetation to grow the way it is, with its multiple species already in place; it did not evolve from anything else, and neither is there any suggestion or indication that there is any relationship between trees and hippos.

On the fifth day, God created the fish and the birds. Again it clearly says "after their kind," meaning that God created the fish and the birds the way they are, with all the reproductive organisms and vital organs already in place, including their own unique DNA codes. Nowhere is there any indication or suggestion that these species evolved from some other being or that they have any relation to the already existing vegetation. God said, "Be fruitful and multiply," meaning that there was also in place the difference between the male and the female species.

"And God said, Let the earth bring forth living creatures after their kind, cattle, and creeping things, and beasts of the earth after their kind: and it was so And God made the beasts of the earth after their kind, and the cattle after their kind, and everything that creepeth upon the ground after its kind: and God saw that it was good" (Genesis 1:24–25). In the beginning of the sixth day, God created the rest of all living things—creatures, cattle, creeping things, and the beasts—again each after their own kind, and again there is no indication or suggestion that there are family ties between one species and another. Obviously, these species were created as they are. Evidence of a relationship between cattle and any other beast or anything that creeps on the ground does not exist. Macroevolution, for this reason, should therefore be considered just another scientific fairytale. This is not to say that microevolution is also unfounded; species can and will still change within their kind as a result of selective breeding or changes in environment. Even mankind itself is more or less indisputable evidence that microevolution does occur. If we examine how we evolved from early primitive creatures to the super intelligent beings we are today, it is obvious that something did happen, that changes did occur. However, it never did happen that we as a species underwent a drastic change—that our tissues or organs changed into a completely different form. Subsequently, there is no way that a mosquito could have evolved into an ape and then, further, into a human being.

Considering the evidence, I believe it would be mere foolishness for a scientist to ignore the involvement of a supernatural God regarding the origin of life on Earth. However, I also believe it would be mere foolishness for a Creationist to ignore the advances of scientific discoveries. There is a perfect harmony between science and belief in God. It is important not to ignore the laws of nature; after all, we must realize it was Almighty God who established these Laws in the first place. And, not ignore what Albert Einstein, the world's greatest thinker said: "I want to know how God thinks, the rest are only details." I believe what Einstein meant by this is, that we should focus our research on how God created life, not trying to reestablish the origins without consideration for the Almighty Creator. We also must not ignore logic and common sense, or reality for that matter. The lack of proof and evidence for the evolutionary theories only make Genesis more credible and make evolution sound like another scientific fairytale. Atheist scientists like Stephen Hawking, Stephen Jay Gould and Richard Dawkins, using their vivid imaginations, propose different ideas for the origins of life; they resort to apes and aliens as an alternative to Devine creation of humanity by Almighty God.

Watching Ben Stein's documentary *Expelled—No Intelligence Allowed*, was an exhilarating experience for me. Ben Stein was brilliant in his ability to manipulate top leaders in the atheist movement into letting down their defenses and confessing what they really believed about God, life on Earth, evolution, etc. One of the most amazing sections of the documentary was Stein's interview with Richard Dawkins. At one point Stein asked Dawkins how he thought life on Earth originated. Dawkins's response was intriguing, because he suggested that perhaps life on Earth had been "seeded" by "extraterrestrials." This admission was equal to a confession that indeed there is an intelligent designer behind all the intricate forms of life, but the designer is definitely not the God of the Bible or any other divine being. Dawkins went on to admit that extraterrestrials, of course, do require a designer as well, but again he provided no room for God.

Stephen Jay Gould, in *Evolution as Fact and Theory*, says, "Einstein's theory of gravitation replaced Newton's, but apples did not suspend themselves in mid-air, pending the outcome. And humans evolved from ape-like ancestors whether they did so by Darwin's proposed mechanism or by some other yet to be discovered." Of course apples did not suspend themselves in midair. Even though Einstein and Newton may have had different ideas about their concepts of gravity, they never did change or alter the laws of gravity. Obviously, Gould insisted that humanity evolved from apes, disregarding all evidence to the contrary. Where Darwin's proposed mechanisms didn't work, he just assumed that there must be other means, perhaps not yet discovered, but again there was no room for God. Gould was saying that he was convinced that humanity evolved from apes, but he had no idea of how or when this occurred.

> And God said, Let us make man in our image, after our likeness: and let them have dominion over the fish of the sea, and over the birds of the heavens, and over the cattle, and over all the earth, and over every creeping thing that creepeth upon the earth. And God created man in his own image, in the image of God created he him; male and female created he them. And God blessed them: and God said unto them, be fruitful, and multiply, and replenish the earth, and subdue it; and have dominion over the fish of the sea, and over the birds of the heavens, and over every living thing that moveth upon the earth. And God said, Behold, I have given you every herb yielding seed, which is upon the face of all the earth, and every tree, in which is the fruit of a tree yielding seed; to you it shall be for food: and to every beast of the earth, and to every bird of the heavens, and to everything that creepeth upon the earth, wherein there is life, *I have given* every green herb for food: and it was so. And God saw everything that he had made, and, behold, it was very good. And there was evening and there was morning, the sixth day. (Genesis 1:26–31)

Toward the end of the sixth day, things become more interesting. This is where God creates mankind. Verse 27 indicates that God made man in His own image, male and female, meaning God made the human species exactly the way they are, with all their vital organs, irreducible complexities, and sexual reproduction system already in place.

"And Jehovah God formed man of the dust of the ground, and breathed into his nostrils the breath of life; and man became a living soul" (Genesis 2:7). God's last act of Creation was where He created man from the dust of the earth. Although the material man was made of was lifeless matter of the earth, it only became a living being when God put life into it. God breathed into man's nostrils the breath of life. The word "breath" is used in various ways in scripture; generally it means "spirit," which seems to be what is meant here. God gave man the spirit of life, meaning the spirit of God himself. When God breathed into man's nostrils, this was precisely the exact point in time when mankind was created. This is where humankind took on a form completely different from that of any other living thing on Earth. This completely elevated man well above any other form of life—even his closest resemblance, such as an ape, monkey, or chimpanzee. By breathing into man's nostrils, God made man more like God Himself, giving man seven very distinct advantages over any other species. By breathing into his nostrils, God gave man a soul; He gave man moral law—the ability to naturally distinguish between what is right and what is wrong—He gave man a level of intelligence, He gave man the power to think and calculate, He gave man the power to create, He gave man the power to reason and make logical decisions, and He gave man the ability to imagine and create things he had never seen before.

Not only that, but He also made man more physically attractive than any other species—even its closest competitor, the ape.

Whenever I happen to look in a mirror, for some reason the image of God always comes to mind; this is probably because I am a believer in God, and it is likely a silent reminder that I was created in His image. I have often wondered what an atheist, such as Richard Dawkins, sees when he looks in his mirror. Most likely some resemblance between himself and one of his early ancestral apes; it is not very likely he sees an image of the God he insists does not exist. Richard Dawkins says, in his own words, "We admit that we are like apes, but we seldom realize that we are apes." Well, perhaps this refers just to Mr. Dawkins's side of the family.

If I ever happen to watch a tornado sweep through a junkyard and I find a Boeing 747 ready for takeoff after the storm dies down, only then will I deviate from my biblical belief that God created all species exactly the way they are and consider some scientific fairytale in which all life started from nothing and for no reason.

The precise biblical details of the origin of life, as well as the universe, may be somewhat vague and without scientific biological explanations; however, the details of how God created life on Earth are much more precise and actually very clear.

Astronomy

Astronomy is considered one of the oldest and most practiced sciences in the history of humankind. It seems that historic cultures have always been actively involved in astronomy. For example, the construction of the Giza pyramids in Egypt several thousands of years ago was undertaken based on astronomical observations; the pyramids geographically represent the Orion constellation. These pyramids are true masterpieces and have rightly earned the title of a wonder of the world. These structures were built with such precision that our current technology is likely incapable of replicating them. Many interesting facts about these pyramids still baffle the minds of archeologists, scientists, and astronomers today.

Do not forget magnificent Stonehenge in southern England. Stonehenge is an ancient monument of huge stones solitarily standing on the Salisbury Plain in Wiltshire, where it has been capturing the imaginations of historians for centuries. Theories about who is responsible for its construction have included the Druids, Greeks, and Phoenicians. Speculation on the reason it was built range from ceremonial purposes to human sacrifice to astronomy. Investigations over the last one hundred years have revealed that Stonehenge was built in several stages from 3000 to 1800 BC. It seems to have been designed to allow for observation of astronomical phenomena—eclipses and summer and winter solstices, for example, which certainly attests to the notion that these early observatories were used to determine the seasons, providing knowledge of when to plant crops and aiding in the understanding of the length of the year.

With the invention of the telescope, astronomy took on a completely different dimension. For the first time in history, astronomers could see past what only the naked eye had been capable of detecting before. They made astonishing discoveries often contradicted by the Catholic Church.

The earliest functional telescopes invented were refracting telescopes that appeared in the Netherlands in 1608. Development of these telescopes is credited to three individuals: Hans Lippershey and Zacharias Janssen, who were spectacle makers in Middelburg, and Jacob Metius of Alkmaar. The States-General of the Netherlands in the Hague discussed the patent applications first of Hans Lippershey, and then of Jacob Metius, on October 2, 1608. Zacharias Janssen was not present at the hearings; at the time he was attending the Frankfurt Fair, where he was trying to sell his extraordinary invention. Lippershey failed to receive a patent, since the same claim for invention had been made by more than one individual, but he was handsomely rewarded by the Dutch government for copies of his design. In June of 1609, world-renowned Italian scientist and mathematician Galileo Galilei became aware of the Dutch inventions, and within a month he built his own instrument. He then greatly improved upon the design in the following year. The invention of the two-element achromatic lens in 1733 improved the quality of refracting telescopes tremendously, correcting color aberrations present in single-element designs and enabling the construction of shorter, more functional telescopes.

The idea that a parabolic mirror could be used as an objective (light-gathering device) instead of a lens was investigated soon after the refracting telescope was invented. The potential advantages of using parabolic mirrors instead of lenses were the reduction of spherical aberration and the lack of chromatic aberration. This resulted in many proposed designs and several attempts to construct such a device. In 1668, Isaac Newton was the first to build a practical reflecting telescope, which still bears his name to this day—the Newtonian reflector. The twentieth century saw the development of a vast variety of telescopes that worked in a wide range of wavelengths, from radio waves to gamma rays. The first specifically designed and constructed radio telescope went into operation in 1937. Since then, a tremendous variety of more complex astronomical instruments have been developed.

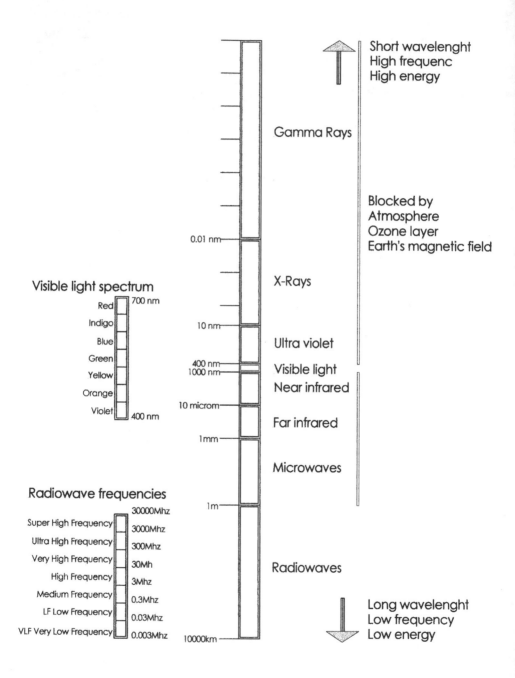

Short wavelenght
High frequenc
High energy

Gamma Rays

Blocked by
Atmosphere
Ozone layer
Earth's magnetic field

0.01 nm

X-Rays

Visible light spectrum

Red — 700 nm
Indigo
Blue
Green
Yellow
Orange
Violet — 400 nm

10 nm

Ultra violet

400 nm
1000 nm — Visible light
Near infrared

10 microm

Far infrared

1mm

Microwaves

Radiowave frequencies

	30000Mhz
Super High Frequency	3000Mhz
Ultra High Frequency	300Mhz
Very High Frequency	30Mh
High Frequency	3Mhz
Medium Frequency	0.3Mhz
LF Low Frequency	0.03Mhz
VLF Very Low Frequency	0.003Mhz

1m

Radiowaves

Long wavelenght
Low frequency
Low energy

10000km

The Electromagnetic Spectrum

Electro Magnetic Radiation is energy transferred by waves of a combination of electrical and magnetic charges, capable of traveling through a vacuum at the universal speed of light. In whatever media radiation is passing through; the speed depends on the media, and is fastest in vacuum. Electro Magnetic Radiation includes radio and microwave signals, infrared (radiant heat), visible light, and ultraviolet, and x-rays and gamma rays. The electromagnetic spectrum extends from below the low frequencies used in radio communication to gamma radiation at the short-wavelength (high-frequency) end thereby covering wavelengths from thousands of kilometers down to the size of an atom. For most of history, visible light was the only known part of the electromagnetic spectrum. The spectrum of visible light may be a very small portion of the electromagnetic spectrum, but consist of many variations of wavelengths. The human eye observes these variations as different colors. The explanation for the color separation is that the refractive index for the various colors is slightly different. The refractive index for red light in glass is 1.513, while the refractive index for violet light in glass is 1.532. This slight difference is enough for the shorter wavelengths of light to be refracted more at a slightly greater angle.

Isaac Newton performed a famous experiment using a triangular block of glass called a prism. He used sunlight shining in through his window to create a spectrum of colors projected on the opposite wall of his room. He realized these were the same colors he had observed earlier in rainbows in the same order-red, orange, yellow, green, blue, indigo and violet, instantly realizing that rainbows are caused by reflection of sunlight in rain droplets at a specific angle. Newton also showed that each of these colors cannot be turned into other colors, but they can recombine to make white light again.

The ancient Greeks recognized that light travels in straight lines and studied some of its properties, including reflection and refraction. Over the years the study of light continued and during the 16th and 17th centuries there were conflicting theories which regarded light as either a wave or a particle.

The first discovery of electromagnetic radiation other than visible light came in 1800, when William Herschel discovered infrared radiation. He studied the temperature of colors by moving a thermometer through light split by a prism. He noticed that the highest temperature was beyond the red. He theorized that this temperature change was due to "calorific rays", which consist of a type of light ray that could not be observed by the human eye. The following year, Johann Ritter worked on the other end of the light spectrum and noticed what he called "chemical rays" (invisible light rays that induced certain chemical reactions) that behaved similar to visible violet rays, but were beyond them in the spectrum, and were later renamed ultraviolet radiation.

During the 1860s, Scottish physicist, James Clerk Maxwell predicted the existence of radio waves; and in 1886, German physicist, Heinrich Rudolph Hertz demonstrated that rapid variations of electric current could be projected into space in the form of radio waves similar to those of light and heat.

In 1895 Wilhelm Röntgen noticed a new type of radiation emitted during an experiment with an evacuated tube subjected to a high voltage. He called these radiations x-rays and found that they were able to travel through parts of the human body but were reflected or stopped by denser matter such as bones. Before long, many uses were found for them in the field of diagnostic medicine. A week after his discovery, Rontgen took an X-ray photograph of his wife's hand which clearly revealed her wedding ring and her bones. The photograph electrified the general public and aroused great scientific interest in the new form of radiation. Röntgen named the new form of radiation X-radiation (X standing for "Unknown"). Hence the term X-rays. Gamma-rays are the most energetic (highest frequency, shortest wavelength) form of electromagnetic radiation. Events relating to supernovae are the primary gamma-ray sources in our galaxy. However, other galaxies, specifically Active Galaxies and Quasars, are known to be sources of gamma-rays as well. Because of the opacity of our atmosphere, gamma-rays usually do not penetrate down to the surface of the Earth; they are blocked by Earth's magnetosphere, ozone-layer, and atmosphere. However, there are techniques that can be employed that search for shockwaves created indirectly by interactions of gamma-rays in our atmosphere.

During the twentieth century, professional astronomy split into two different categories: observational and theoretical astronomy. These two separate fields complemented each other extremely well, with theoretical astronomy explaining observational results and telescopic observations confirming mathematical calculations. Significant advances in astronomy were also made with the invention and introduction of newer technologies, such as the spectroscope and photography. In 1814, German astronomer Joseph von Fraunhofer (1787–1826), making use of his newly invented spectroscope, discovered about six hundred different bands in the spectrum of sunlight. In 1859, Gustav Kirchhoff (1824–1887) explained that these bands represent different elements. It also proved that distant stars were very similar to our own sun, except that they exhibited a wide variety of temperatures, masses, and sizes.

A spectroscope is an instrument that astronomers use to measure the properties of light over a specific portion of the electromagnetic spectrum, typically to identify different materials and elements. It produces spectral lines and measurers their wavelengths over a wide range, from gamma rays and X-rays into the far infrared. With the invention of photographic films, spectrography became much more precise and accurate, utilizing a camera instead of a simple viewing tube. The spectroscope has become the most important scientific tool for the identification and analysis of the compositions of unknown substances, and for studying astronomical phenomena and for the testing of astronomical theories. Hubble's law and the Hubble sequence were all made possible using spectrographs utilizing photographic paper.

Radio astronomy is the study of heavenly bodies that emit radio waves with wavelengths greater than approximately 1 mm. The initial detection of radio waves coming from astronomical objects was made in the 1930s, when Karl Jansky (1905–1950), an American physicist and radio engineer, observed radiation coming from somewhere in the Milky Way. Subsequent observations identified a vast number of different sources of radio emissions, including stars and galaxies, as well as entirely new classes of objects, such as radio galaxies, quasars, and pulsars. The discovery of the cosmic microwave background radiation by Penzias and Wilson, which provided compelling evidence for the big bang theory, was also made through the science of radio astronomy. Radio astronomy is executed by the use of a large radio antenna called a radio telescope, which is used either as a single unit or as several units linked together, employing the techniques of radio interferometry and aperture combination. The use of interferometry permits radio astronomy to obtain higher angular resolutions, as the resolving power of an interferometer is determined by the distance between antennas rather than by their sizes.

Infrared astronomy is the detection and study of the infrared radiation (heat energy) emitted from distant objects in the universe. Astronomical objects in the universe provide an unparalleled amount of information by transmitting tremendous amounts of electromagnetic radiation (including light). Much of this information is in the infrared wavelength, which cannot be seen with the naked eye or with optical telescopes. Infrared astronomy involves the study of just about everything in the entire universe. In the field of astronomy, the infrared region lies within the range of sensitivity of infrared detectors, which is between wavelengths of about 1 and 300 microns (a micron is one millionth of a meter). The human eye detects only 1 percent of light at 0.69 microns and 0.01 percent at 0.75 microns, and so effectively cannot see wavelengths longer than about 0.75 microns unless the light source is extremely bright.

Only a small amount of this infrared information reaches Earth's surface, because most of it is absorbed by our atmosphere. Yet by studying this small range of infrared wavelengths, astronomers have uncovered a wealth of new information. Only since the early 1980s have we been able to send infrared telescopes into orbit around Earth, well above the atmosphere, which hides most of the universe's infrared light from us. The new discoveries made by these infrared satellite missions have been astounding. The first of these satellites, the Infrared Astronomical Satellite (IRAS), detected about 350,000 new infrared sources, increasing the number of cataloged astronomical sources by about 70 percent.

Gamma ray astronomy is the study of astronomical objects at the shortest wavelengths of the electromagnetic spectrum. Gamma rays cannot be seen with the naked eye but can only be observed directly by satellites, such as the Compton Gamma Ray Observatory, or by specialized telescopes called atmospheric Cherenkov telescopes.

Amateur astronomers have contributed a lot to important discoveries and still play an active role, especially in the discovery and observation of moving objects in the universe.

The Speed of Light

In 1638, Galileo was the first known scientist to conduct an experiment in an attempt to measure the speed of light by observing the delay between uncovering a lantern and its perception some distance away. He arranged two observers, each having a lantern equipped with shutters, at a mile distance from one another. The first observer would open the shutters of his lamp, and the second observer, upon seeing the light, would immediately open the shutters of his lamp. Galileo was unable to distinguish whether light travel was instantaneous or not, but he did conclude that if it weren't, it must nevertheless be extraordinarily fast. Based on the modern value of the speed of light, the actual delay in this experiment would be about eleven microseconds, which is impossible to measure without the use of sophisticated equipment not available in 1638.

The speed of light, officially denoted by the letter c, is a physical constant, so named because it is the speed at which light and all other electromagnetic radiation, such as radio waves, travel in a vacuum. Its value is exactly 299,792,458 m/s as accepted by the International Bureau of Weights and Measures. The speed of light is generally approximated as 300,000 km, or 186,000 mi, per second, meaning that light will travel around Earth almost eight times in just one second. In Einstein's theory of relativity, c connects space and time in space-time, and it appears in Einstein's famous equation of mass–energy equivalence, $E=mc^2$, the equation everybody knows about though very few understand what it actually means. $E=mc^2$ means that energy and mass are equivalent and transmutable (changeable from one form to another). The mathematics of general relativity appeared to be inconceivably difficult; consequently, only a small number of people in the world could fully understand the theory in precise detail. This remained the case for about forty years, until about 1960, when a critical revival in interest transpired, making general relativity central to physics and astronomy. Relativity simply explains theories of physics proposed by Albert Einstein. Einstein theorized, among other things, that space and time cannot be considered separate entities and that nothing can move faster than the speed of light. The perception of space-time is different for an observer standing still on Earth than it is for someone moving very fast away from or toward it. What we see is relative to (depends on) our acceleration as we move. For example, if we are travelling in an automobile on the highway at a speed of 50 kmh and another automobile approaches us from behind at a speed of 100 kmh, the relative speed of the second automobile would be 50 kmh. However, when another automobile approaches from the front at 100 kmh, the relative speed of that automobile would be 150 kmh.

The speed at which light travels through transparent materials, such as glass or air, is less than c. The ratio between c and v, the speed at which light travels in a material, is called the refractive index and is represented by the letter n ($n=c/v$ or inversely $v=c/n$). For example, the refractive index of visible light in glass is typically around 1.5, meaning that light in glass travels at $c/1.5$, which is equal to 200,000 km/s. The refractive index of air for visible light is about 1.0003, so the speed of light in air is a little less than, but very close to, c. Denser materials, such as water, glass, and diamond, have refractive indexes of around 1.3, 1.5, and 2.4, respectively, for visible light.

Another consequence of the finite speed of light and other electromagnetic waves, such as radio transmissions, is that communications between Earth and spacecraft are not instantaneous. There is a brief delay from the transmitter to the receiver, which, of course, becomes more noticeable as the distance increases. This delay was significant for communications between ground control in Houston, Texas, and the crew of Apollo 8, the first manned spacecraft to orbit the moon. For every question, ground control had to wait at least three seconds for the answer to arrive. The communications delay between Earth and the distant planet Mars is almost ten minutes. As a consequence of this, if a robot probing the surface of Mars were to encounter a problem, its human controllers here on Earth would not be aware of this dilemma until ten minutes later; it would then take at least another ten minutes for instructions to travel from Earth back to Mars to correct the problem.

The first calculated estimate of the speed of light was made in 1676 by Danish astronomer Ole Christensen Rømer (1644–1710). Rømer noticed that the periods of Jupiter's innermost moon, Io, appeared to be shorter when Earth was approaching Jupiter than when Earth was receding from Jupiter. He consequently concluded that light travels at a fixed speed, and he estimated that it would take light twenty-two minutes to cross the diameter of Earth's orbit. Dutch scientist and astronomer Christiaan Huygens (1629–1695) combined these observations with an estimate for the diameter of the Earth's orbit to calculate an estimate of the speed of light of 220,000 km/s, only 26 percent lower than today's precisely calculated value.

In his 1704 book *Opticks*, British physicist, mathematician, and astronomer Isaac Newton (1643–1727) also studied Rømer's calculations of the finite speed of light and calculated that it would take about "seven or eight minutes" for the light of the sun to reach Earth, actually not far from today's value of eight minutes and nineteen seconds. Newton also questioned Rømer as to whether eclipse shadows were of different colors; after finding out that they weren't, he correctly concluded that all colors of the light spectrum travel at the same speed.

Distance in Space

Determining distance is assigning a numerical value to how far objects are apart. Logic tells us that the shortest distance between two points is a straight line.

Here on Earth we all have grown up with familiar units of distance, namely millimeters, centimeters, meters, and kilometers, or in some parts of the world with statue miles, one mile equaling 1.609344 km. We consider a millimeter to be a very tiny unit compared to a kilometer. However, when we consider the vast distances in space, we find that earthly units are completely inadequate to express these distances. To use the same units as we use here on Earth we would have to apply extremely large numbers, which would be similar to expressing the distance between Amsterdam and Los Angeles in millimeters (89,630,000,000, or 89.63 billion) instead of the more common units of kilometers (8,963 km).

In astronomic circles, a total different system of expressing distances is utilized. The most common is the light-year (ly), which represents the distance light travels in a one-year period. Light travels at a finite speed of 299,792,458 m/s, which equals 9,460,730,473,000, or 9.461 trillion km per year. This means that light travels the distance from the moon to Earth (384,400 km) in just 1.28 seconds, and it travels the distance from the sun to Earth (149,597,870 km) in a mere 8.3 minutes. So instead of saying that the sun is almost 150 million km away, we simply state that the sun is just 8.3 light minutes away. When we visually observe the sun, we are not looking at the place where the sun is now, but where the sun was 8.3 minutes ago.

Now, when we consider our nearest star, Alpha Centauri, its light needs to travel 4 years and 4 months in order to reach us; thus Alpha Centauri is 4.3 ly from Earth. In kilometers this translates to the incomprehensible number of about 40,681,156,000,000 (40.681 trillion) km. These gigantic numbers start to make sense only when we compare them to the US national debt. If we were able to fly a commercial jet airplane to Alpha Centauri, our closest neighboring star at only 4.3 ly away, at a constant speed of 1,000 kms, it would take us approximately 4.64 million years to reach our destination. Considering that our neighborhood of the Milky Way Galaxy is about 100,000 ly in diameter, this gives us a faint perception of how vast our galaxy really is and how petite we as humans really are.

For an intermediate way of expressing distances in space, astronomers work with a unit called an astronomical unit (AU), which is mainly used to express distances within our own solar system. An AU is considered the distance from Earth to the sun, approximately 150 million km. Today, astronomers have accepted that the AU is precisely 23,455 times the radius of Earth. To be more precise, it is defined by the International Bureau of Weights and Measures as 149,597,870.7 km.

A third, however less common, way to express distances in space is the parsec. This unit of measurement is based on geometry rather than speed and time. It is defined as the length of the adjacent side of an imaginary triangle in space. The two dimensions that specify this triangle are the parallax angle (1 arc second) and the opposite side (1 AU). Given these two measurements, using the rules of trigonometry, the length of the adjacent side (the parsec) can be calculated. The official distance of a parsec equals $3.08568025 \times 10^{16}$ m. Wow! To express this in simpler terms, it is equivalent to about 3.26156 ly, or just under 31 trillion km.

Outer space, more simply called space, refers to the relatively empty regions of the universe outside the atmospheres of celestial bodies. There is no distinct boundary between the Earth's atmosphere and space, as the density of the atmosphere gradually decreases with altitude. Space within the solar system is called interplanetary space; outside our solar system, it is referred to as interstellar space. Outer space is certainly spacious, but it is not exactly empty; it is sparsely filled with several types of organic molecules (as discovered by microwave spectroscopy), cosmic background radiation left over from the supposed big bang, and cosmic rays, which include ionized atomic nuclei and various subatomic particles. There are also some gasses, plasma, and dust present, as well as small meteors. Additionally, there are remnants of human life in outer space, such as materials and debris left over from previous manned and unmanned spacecraft, which have become a potential hazard to today's spacecraft and the thousands of satellites orbiting Earth. Some of this debris reenters the atmosphere periodically. Although Earth is currently the only known body within the solar system—or, for that matter, the entire universe—to support life, current evidence suggests that in the distant past the planet Mars may have possessed bodies of liquid water on its surface. Therefore Mars, during a brief period in its history, may have been capable of forming life. At present, though, most of the water remaining on Mars is frozen. If life does exist at all on Mars, it is most likely to be located underground, where liquid water can still exist.

Astronomers employ two different systems of measuring the brightness of heavenly bodies in the night sky. Absolute magnitude is a measurement of the true brightness of a star as if all stars were the same distance (32.6 ly) from the observer. Apparent magnitude, a more common measurement, is a measure of brightness of celestial bodies as seen by an observer on Earth. The brighter the object appears, the lower the value of its magnitude. Our full moon, for example, has an apparent magnitude of -12.6 while Sirius, the brightest star in the celestial sphere, has an apparent magnitude of -1.4.

To try to gain comprehension of how enormous things in the universe really are, we are going to compare our sun to four other stars in our galaxy. First, Alpha Centauri, because it happens to be our closest neighbor at 4.3 ly away, then Sirius, because it is the brightest star in the night skies at 8.6 ly away. Then we will explore Antares, at 600 ly from us, because of its gigantic size. Finally, we will look at Canis Majoris, 4900 ly from us, because this is absolutely the biggest heavenly body that astronomers have been able to detect thus far; in fact, it is so big it will absolutely boggle your mind!

In order to gain some understanding of how vast distances in our universe really are, I will introduce a distance scale that I simply refer to as the alpha scale. Here I scale the diameter of Earth's orbit around the sun, 299,195,741 km, down to a line only 10 mm long, or about half the size of a dime. This means that 10 mm in our Alpha scale represents about 300 million km. At this scale, our Milky Way Galaxy, with a diameter of 100,000 ly, would translate to a line measuring about 31,620 km long, more than two and a half times the diameter of Earth! At this scale, Earth itself would be represented as being smaller than an atom; the most powerful microscope would be needed to find it. In fact, it would be ten million times easier to find a tiny little needle in a humongous haystack than it would be to find the location of Earth in the Milky Way Galaxy.

Alpha Centauri, located in the southern constellation Centaurus, has an apparent magnitude of -0.01, and its size is about 1.227 times that of our sun, or 1,707,984 km in diameter, and it is our closest neighboring star at 4.3 ly away. With the naked eye, Alpha Centauri appears to be a single star, but it is actually a binary system, meaning it consists of two related stars orbiting each other. The secondary star, 0.865 times the size of our sun, is calculated to measure 1,204,080 km in diameter. Translating the distance of 4.3 ly to our Alpha scale, with 10 mm representing about 300 million km, Alpha Centauri would be located about 1.36 km from Earth.

Sirius is located in the constellation Canis Major and is the brightest star in the night sky. Only the moon, Venus, Jupiter, and Mars are brighter. With an apparent magnitude of -1.46, it is almost twice as bright as Canopus, the next brightest star. With the naked eye, Sirius also appears to be a single star, but it is a binary system as well. Sirius, with a diameter of 2,465,232 km, or 1,711 times the size of our sun, appears bright because of both its intrinsic brightness and its close distance to Earth. At a distance of 8.6 ly away, Sirius is our second-closest neighboring star and is about twice as massive as our sun. In 1868, Sirius became the first star to have its velocity measured. Sir William Huggins examined the spectrum of this star and observed a noticeable red shift, leading him to conclude that Sirius was receding from our solar system at about 40 km/s. Astronomers later calculated the actual value to be -7.6 km/s. Mr. Huggins not only overestimated the velocity, but he also had the direction wrong; the minus sign denotes that Sirius is approaching our solar system instead of moving away from us. This, of course, does not mean that Sirius is on a collision course with our solar system, and even if it were, there would be nothing to worry about, as it would take almost four hundred thousand years for it to reach our solar system at the velocity it is traveling.

The alpha scale puts Sirius 2.751 km away from Earth.

Antares is a red super giant star in our Milky Way galaxy, located in the constellation Scorpius, and is the sixteenth brightest star in the nighttime sky. It is one of the four brightest stars near the ecliptic, the others being Aldebaran, Spica, and Regulus. Antares is a slow variable star with an average apparent magnitude of +1.09. Antares is located about 600 ly from our solar system, or 189,723 km on the alpha scale. The enormous size of Antares commands special attention; with a diameter of 1.1136 billion km, it is not 3.72 times the size of our sun, but it is 3.72 times the size of Earth's orbit around the sun. On the alpha scale, Earth's orbit around the sun represents a line of 10 mm; Antares's diameter represents a line 37.22 mm long.

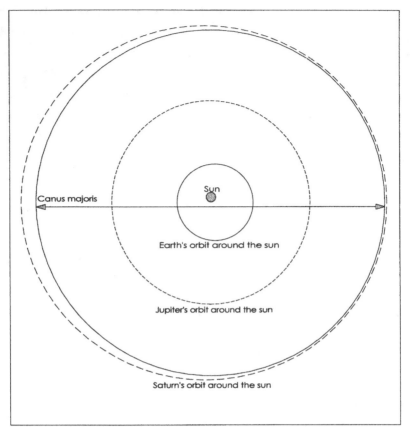

The size of Canis Majoris compared to the orbit of Saturn

VY Canis Majoris is a supergiant red star in our Milky Way Galaxy, located in the constellation Canis Major. With a diameter of 2,900,000,000 (2.9 billion) km, Canis Majoris is the absolute biggest known heavenly body in the universe, and with an apparent magnitude of 7.9607, it is also one of the most luminous. In English, the Latin name Canis Majoris means "big dog," and that is exactly what Canis Majoris is. As a matter of fact, it is the biggest star that we know of, located about 4,900 ly away from Earth. This star is so big it will blow your mind! Unlike most supergiant stars, which occur in either binary or multiple star systems, Canis Majoris is a single star. Visual observations in 1957 and high-resolution imaging in 1998 showed that Canis Majoris does not have a companion star. Our Alpha Scale puts

Canis Majoris at a distance of 1,549.4 km away from us. Its diameter is represented on the alpha scale as a line 96.93 mm long, almost ten times Earth's orbit around the sun. It would take 7×10^{15} (seven quadrillion) Earths to fill the volume of Canis Majoris.

To summarize our four-star comparison, we visualize a gigantic circle with a diameter of 31,600 km, almost three times the diameter of Earth, representing the Milky Way Galaxy. Somewhere within this gigantic circle we place another tiny circle with a diameter of only 10 mm, about half the size of a dime. This tiny 10 mm circle represents Earth's orbit around the sun. From our 10 mm circle, we place a dot 1.3 km away; this would be the relative location of Alpha Centauri. Sirius's dot would be 2.75 km away, and Antares would be 190 km away, represented by a circle of 37.2 mm, while Canis Majoris would be located 1,549 km away, represented by a circle of almost 97 mm. These four stars are still our closest neighbors, and so far we have ventured into outer space by a distance less than 5 percent of the diameter of our Milky Way Galaxy. If we were to modify the alpha scale so that only Earth, not Earth's orbit around the sun, were to be represented as a disk 10 mm in diameter, our sun would be represented as a disc with a diameter of 1.09 m at a distance of 117 m away. Canis Majoris's disc would have a diameter of approximately 2.3 km and would be located some 3.6 million km away. Wow!

A galaxy is a massive, gravitationally bound system of stars and stellar remnants containing an interstellar medium of gas and dust, and an important but poorly understood component tentatively dubbed "dark matter." Dark matter is something nobody understands; nobody knows what it is, and nobody has ever seen or observed it. However, dark matter is required to support the big bang theory, and therefore scientists feel that it must be there.

In 1610, Galileo, using a telescope to study the Milky Way, discovered that our galaxy consists of a humongous number of faint stars. In 1750, Thomas Wright, in his *An Original Theory or New Hypothesis of the Universe*, speculated correctly that a galaxy might be a rotating body of a huge number of stars bound together by gravitational forces—just like our solar system, except on a much larger scale. The Hubble Space Telescope, launched in 1990, photographed an area of space equal to a grain of salt held at arm's length. After a long exposure of this relatively empty part of the universe, it produced a stunning image, providing evidence that there are about 125 billion (1.25×10^{11}) galaxies in the universe. Most of these galaxies measure somewhere between 30,000 to 350,000 ly in diameter and are usually separated from each other by millions of light-years.

Intergalactic space (the space between galaxies) is not exactly completely empty space; it is filled with a tenuous gas of an average density of less than one atom per cubic meter.

Typical galaxies range from dwarfs with as few as ten million stars up to giants with perhaps a hundred trillion stars; all these stars orbit the galaxy's center of mass. Galaxies may contain many multiple-star systems, star clusters, and various interstellar clouds. A common form is the elliptical galaxy, which has an ellipse-shaped light profile. Spiral galaxies are disk-shaped groupings with dusty, curving arms. Galaxies with irregular or unusual shapes are known as irregular galaxies and are caused by the gravitational pull of neighboring galaxies. Such interactions between nearby galaxies, which may ultimately result in galaxies merging, may induce episodes of significantly increased star formation, producing what is called a starburst galaxy. As an example of such an interaction, the Milky Way Galaxy and the nearby Andromeda Galaxy are moving toward each other at about 130 km/s, and depending upon the lateral movements, these two galaxies could possibly collide in about 5.8 billion years. Not long ago I shared this information in casual conversation with a colleague. Suddenly he became quite worried and concerned, so I inquired what the problem was. "How long, you say?" he asked. I said, "5.8 billion years." "Oh, thank God," he replied, "I thought you said 5.8 million!"

Observational data suggests that super massive black holes may exist at the center of many, if not all, galaxies. A black hole, according to the general theory of relativity, is a region of space from which nothing, not even light, can escape. It is the result of the deformation of space-time caused by a very compact mass. Around a black hole is an undetectable surface that marks the point of no return; this is called an event horizon. The word "black" is used in the name because a black hole absorbs all the light that comes near it, reflecting nothing, just like a perfect black body in thermodynamics. Our own Milky Way Galaxy appears to harbor at least one such black hole within its nucleus.

The Andromeda Galaxy is a spiral galaxy approximately 2.5 million ly from us in the constellation Andromeda, with a diameter at its widest point of 141,000 ly. It is also known as Messier 31, M31, or NGC 224. Andromeda is the nearest spiral galaxy to our own Milky Way, but it is not the closest galaxy overall. It is visible as a faint smudge on a moonless night and is one of the farthest objects visible to the naked eye; it can even be seen from urban areas with a simple pair of binoculars. It gets its name from the area of the sky in which it appears—the area of the Andromeda constellation. Andromeda is the largest galaxy of the Local Group, which consists of Andromeda, the Milky Way, the Triangulum, and about thirty other smaller galaxies. The 2006 observations by the Spitzer Space Telescope revealed that M31 contains about one trillion (10^{12}) stars, almost ten times more than the number of stars in our own galaxy, which is estimated to be100–130 billion. In 2006, it was estimated that the mass of the Milky Way was about 80 percent of the mass of Andromeda; however, in 2009 further studies concluded that Andromeda and the Milky Way are about equal in mass. At an apparent magnitude of 3.4, the Andromeda Galaxy is notable for being one of the brightest Messier objects, making it easily visible to the naked eye even when viewed from areas with moderate light pollution. Although it appears more than six times as wide as the full moon when photographed through a larger telescope, only the brighter central region is visible with the naked eye.

The Sombrero Galaxy (also known as M 104 or NGC 4594) is an unbarred spiral galaxy in the constellation Virgo. It has a bright nucleus, an unusually large central bulge, and a prominent dust lane in its inclined disk. The dark dust lane and the bulge give this galaxy the appearance of a sombrero. The Sombrero galaxy has an apparent magnitude of +9.0, making it easily visible with amateur telescopes. The large bulge, the central super massive black hole, and the dust lane all attract the attention of professional astronomers. The Sombrero Galaxy is located approximately 29.3 million ly away from us.

The Whirlpool Galaxy (M51a, or NGC 5194), located within the constellation Canes Venatici, 3.5° southeast of the easternmost star of the Big Dipper, is an interacting grand-design Spiral galaxy located at a distance of approximately 31 million ly away. It is one of the most beautiful and famous galaxies in the sky. In his fascinating DVD *Indescribable*, Louie Giglio refers to this galaxy as the "darling" of astronomy because it is the absolute most beautiful sight ever to be observed in the night sky, because its location is perpendicular to our plane of sight. The distance across the Whirlpool Galaxy is about 90,000 ly.

The Whirlpool Galaxy and its companion (NGC 5195) can easily be observed by amateur astronomers, and the two galaxies may even be seen with binoculars. When seen through a 100 mm telescope, the basic outlines of M51 and its companion are quite visible. Under dark skies and with a moderate eyepiece using a 150 mm telescope, M51's intrinsic spiral structure can easily be detected. The Whirlpool Galaxy is also a popular target for professional astronomers, who study it to further their understanding of galactic structure (particularly structure associated with the spiral arms).

In 1992, the orbiting Hubble Space Telescope spotted a peculiar formation, a very distinct image of a cross, in the center of the Whirlpool Galaxy. The startling photographic image was transmitted back to Earth, released during the second week of June 1992, and published in the June 22 edition of *TIME* magazine. Astronomer H. Ford interpreted the image as a dark *X*, consisting of two clouds of dust, silhouetted across the black hole in the galaxy's nucleus. The dust ring stands almost perpendicular to the relatively flat spiral nebula. A secondary ring crosses the primary ring on a different axis—a phenomenon that cannot be explained. Another phenomenon that cannot be explained is why these dust rings are there in the first place; a black hole absorbs absolutely everything in its surroundings—all matter, even light.

Image by NASA Hubble telescope

Jesus on the Cross

Louie Giglio produced a fascinating series of DVDs called the Passion Tale Series. This series presents an incredible visual seminar on our galaxy. In his DVD *Indescribable*, Louie also points to the cross in the Whirlpool Galaxy, describing the greatness and the glory and the majesty of our creator. Giglio uses Psalm 33:6 ("By the word of Jehovah were the heavens made, and all the host of them by the breath of his mouth") to attempt to describe just how big our God is compared to the universe.

Not only does Louie talk about how inconceivably big our God is and the significance of His placement of His signature in the Whirlpool Galaxy thirty-one million light-years away, but he goes on to speak of how this "universe-creating" God also weaved our bodies together. He goes on to talk about laminin. Although laminin is a biological phenomenon rather than astronomical, Louie points out that laminin also contains the signature of Almighty God. Laminin is defined by the Webster Medical Dictionary as a "glycoprotein that is a component of connective tissue basement membrane and that promotes cell adhesion." In other words, laminin are cell-adhesion molecules—molecules that "glue" our bodies together. Just as gravity or atoms or molecules are the glue that holds the physical world together, laminin glycoprotein is one example of this form. The graphical representation of laminin, the form of a cross, clearly illustrates how God's design is evident. Without laminin, our bodies would literally fall apart! Louie shows an image of what laminin looks like. The structure of laminin resembles the image of a cross. This image is not a Christian interpretation of laminin; this is the exact image found in any medical or scientific literature. Louie related laminin to Paul's words in Colossians 1:15–17: "Who is the image of the invisible God, the firstborn of all creation; for in him were all things created, in the heavens and upon the earth, things visible and things invisible, whether thrones or dominions or principalities or powers; all things have been created through him, and unto him; and he is before all things, and in him all things consist."

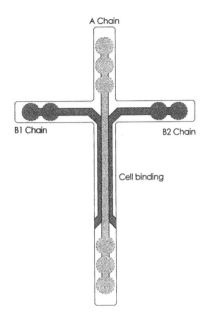

A Chain

B1 Chain

B2 Chain

Cell binding

Structure of Laminin compared to the shape of the Cross

I believe it would be mere foolishness to accept these images as physical proof that God created everything, I believe these images serve as a silent reminder that God did create everything. However, anyone can interpret these images in whatever way they feel comfortable with. Personally, I believe that the images of the cross at the core of the Whirlpool Galaxy and in the structure of laminin are nothing more than the signature of Almighty God Himself. This is Almighty God showing off His splendor, His greatness, His glory and majesty, simply saying to us, "Look what I made for you and how I made you." The world's most powerful telescope is required to find God's signature in the farthest reaches of the universe, while the world's most powerful microscope is needed to find His signature in the minutest parts of our bodies. Not only does Almighty God show us what He made for us, but He also shows us why He made it. The symbol of the cross reminds us of the dying, suffering, shame, and humiliation that Christ endured for the sake of sinful and disobedient humanity in desperate need of redemption. The cross is one of the most ancient human symbols and is used by many religions, mainly Christianity. I am certainly not insinuating that I know how God thinks, but I do get the feeling that by leaving His signature at the core of the Whirlpool Galaxy, God is reminding us of how small and petite we as human beings really are and how humongous we are compared the minutest details of His creation.

Giglio's DVD collection is absolutely a positive addition to the library of any Christian with an interest in astronomy and science.

Our Sun is the star at the center of our solar system, located in the Milky Way Galaxy, and it is one of an estimated 100 to 130 billion stars in our galaxy. It has a diameter of about 1,392,000 km (865,000 mi), about 109 times the size of Earth, and its mass is about 2×10^{30} kilograms (330,000 times that of Earth) and accounts for about 99.86 percent of the total mass of our entire solar system. About three-quarters of the sun's mass consists of hydrogen, while the rest is mostly helium. Less than 2 percent consists of heavier elements, including oxygen, carbon, neon, iron, and a few others.

The color of the sun is white, although from the surface of the Earth it appears to be yellowish; the color differential is caused by atmospheric diffusion. Its spectral class label is G2, indicating that the sun's surface temperature is approximately 5,505°C or 9,941°F. At one time our sun was regarded by astronomers as a fairly small and relatively insignificant star; now our sun is presumed to be brighter than almost 85 percent of all the stars in our galaxy, most of which are red dwarfs. The absolute magnitude of the sun is +4.83; however, as the closest star to Earth, the sun is the brightest object in the sky, with an apparent magnitude of -26.74. The sun's hot corona continuously expands into space, creating the solar wind, a hypersonic stream of charged particles that extends to the heliopause at roughly 100 AU.

The sun travels through the Local Interstellar Cloud in the Local Bubble zone, within the inner rim of the Orion Arm of the Milky Way, which is a quiet region of the galaxy. Within the star-forming regions of our galaxy, such as the Orion Arm or the core of our galaxy, things are a lot more chaotic, with extreme temperatures. These are areas where we really would not want to be. Of the fifty nearest stellar systems within 17 ly from Earth, the sun ranks fourth in mass. The solar system orbits the center of the Milky Way at a distance of approximately 24,000–26,000 ly from the galactic center, or core, completing one clockwise orbit, as viewed from the galactic North Pole, about every 225–250 million years. The mean distance of the sun from Earth is approximately 149.6 million km, though this varies as Earth moves from perihelion (farthest away from the sun) in January to aphelion (closest to the sun) in July.

The energy of sunlight supports almost all life on Earth by photosynthesis (a process converting carbon dioxide into organic compounds, using energy from sunlight) and is the driving force of all Earth's climate and weather. The enormous effect of the sun on the Earth has been recognized since prehistoric times, and the sun has been regarded by some cultures as a god. An accurate scientific understanding of the sun developed rather slowly, and as recently as the nineteenth century prominent scientists still had little knowledge of the sun's physical composition and source of energy. This understanding is still developing; there are a number of inconsistencies in the sun's behavior that still need to be explained. Ernest Rutherford suggested that the sun's output could be maintained by an internal source of heat and suggested radioactive decay as the source. However, it was Albert Einstein who would provide the essential clue to the source of the sun's energy output with his mass-energy equivalence relation $E=mc^2$.

A supernova is a stellar explosion that creates an extremely bright object, initially made of plasma—an ionized form of matter. A supernova may, for a period of time, be much brighter than its host galaxy before fading from view over several weeks or months. During this brief period of time, the supernova radiates as much energy as the sun would emit over about ten billion years. The explosion expels much or all of a star's matter at a velocity of up to a tenth the speed of light, driving a shock wave into the surrounding interstellar gas. This shock wave sweeps up an expanding shell of gas and dust called a supernova remnant.

There are many explanations to describe a quasar. Quasars are extremely luminous and were first identified as being high redshift sources of electromagnetic energy, including radio waves and visible light, that were point-like, similar to stars, rather than extended sources similar to galaxies. Shortly after quasars were discovered, scientists began to study and observe these objects and discovered something very unusual. They observed spectral emission lines where some quasars had been previously, but they could not identify exactly what those lines were. In the 1960s, after careful observation, scientists concluded that those emission lines originated from the quasars themselves. The quasars, scientists proposed, had undergone extremely large redshifts. This resulted in the idea that the quasars were traveling at enormously high velocities, up to 93 percent of the speed of light. Scientists concluded that this provided evidence for a constantly expanding universe.

A quasar is understood to be an extremely bright and distant active nucleus of a young galaxy. Quasars were first identified as being high-redshift (thus very distant) sources of electromagnetic energy, including radio waves and visible light, which were much similar to stars rather than galaxies. However, in his book *Seeing Red*, Halton Arp clearly shows that these supposedly very distant objects are actually part of much closer galaxies, suggesting that their high redshift is not related to distance and thus contradicting the standard big bang theory. His work is now suppressed in the United States (he was banned from using their telescopes), and he later worked at the Max Planck institute in West Germany. This is a sad insight into the state of modern physics and its suppression of the truth to maintain existing beliefs over observed facts.

Scientists believe that our solar system formed from a giant cloud of dust and gas. Almost a century prior to the discovery of the first asteroid, comets were realized to be phenomena that occur outside of Earth's atmosphere; comets are small balls of rock and ice that orbit the sun just like everything else in the solar system. Our understanding of the local region of our galaxy has changed significantly over the past few decades, and it continues to change even today. Our scientifically proven understanding of the solar system has one central star, eight main planets and their associated moons, a large number of asteroids between Mars and Jupiter, and more asteroids located throughout the solar system, and then a gradually increasing disk of asteroids and comets beyond the orbit of Pluto referred to as the Kuiper Belt.

Our solar system is just another of God's incredible creations, consisting of, among other things, our sun, orbited by eight celestial bodies, called planets, bound together by mysterious gravitational forces. The four smaller inner planets—Mercury, Venus, Earth, and Mars—are called the terrestrial planets; they are primarily composed of rock and metal. The four outer planets—Jupiter, Saturn, Uranus, and Neptune— are called the gas giants because they are composed mainly of hydrogen and helium and are far more massive than the terrestrial planets.

The thousands of objects in the asteroid belt that lies between Mars and Jupiter are very similar to the terrestrial planets; they are also composed mainly of rock and metal.

For thousands of years, humanity was not aware of the existence of the solar system. People always believed that Earth was flat, stationary, and the center of the universe. Nicolaus Copernicus was the first astronomer to develop a mathematically predicted heliocentric system. His seventeenth-century successors, Galileo and Johannes Kepler, advanced the science of physics, leading to the gradual acceptance of the idea that Earth orbits around the sun and that all the planets are governed by the same physical laws that govern the movements of Earth. Every planet moves around the sun not in a perfect circle but in an elliptical orbit. The closest distance of a planet to the sun is called perihelion, while the planet's farthest distance from the sun is called aphelion.

A centrifugal force is experienced when an object is moving in a circular path and seems to push away from its center. Its magnitude, or force (F), is represented by the formula $F=mv^2/r$ where m is the mass of the object (expressed in kilograms), v is the speed of the object (expressed in seconds), and r is the radius of its path (expressed in meters). If you were to attach a 10 kg object to the end of a 2 m rope and swing it overhead in a circular motion, you would create a centrifugal force, where you were the center of rotation. The length of the rope would determine the radius of distance, and the more force you applied, the faster the object would move, and the stronger the rope would need to be. If you were to change the rope to, for instance, an elastic bungee cord, you would find that the faster the object moved, the more the bungee cord would stretch, and the distance of the object would increase from the center of rotation. A similar phenomenon occurs, for example, when you swing a bucket filled with water in a vertical loop. The water stays in the bucket because the centrifugal force pushes it to the bottom of the bucket. The more you accelerate the rotation, the faster the bucket moves, and the more stress is applied to the bottom of the bucket. Our solar system works on the exact same principles, except you are replaced by the sun as the center of rotation, the rope is replaced by the sun's gravitational force, and the object is replaced by a planet— for example, Earth. Our solar system is designed and created in a very specific way; it is bound together by a combination of gravitational forces, mass, distance, velocity, and centrifugal forces. Every one of these forces is precisely calculated, balanced, and harmonized to make the existence and function of our solar system possible.

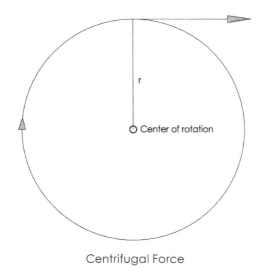

Centrifugal Force

The combination of the sun's gravitational force, the orbital speed of Earth (29.78 km/s, or 107,200 kmh), Earth's mass (5.9736 × 10²¹ tons), and Earth's precise distance from the sun (149,597,870 km) keeps Earth in a perfectly balanced orbit around the sun. If any of these parameters were to change—for example, if Earth were to move at a slower speed— it would undermine the centrifugal force, the sun's gravitational force would pull Earth toward the sun at an accelerated rate, and finally the sun would absorb Earth completely. On the other hand, if Earth were to move any faster, it would gradually extend its orbit and ultimately disappear into outer space. The exact same principle applies to all other planets in our solar system as well.

Centrifugal forces pushing outward, and gravity pulling inward, keep Earth in perfect orbit around the sun. This amazing balance of mass, distance, speed, and gravity clearly imply perfect design.

Johannes Kepler (1517–1630) was a German astronomer and mathematician of the late sixteenth and early seventeenth centuries. Kepler developed a different perception of the solar system; he didn't believe planetary orbits were necessarily circles but could instead be closed curves, such as ellipses or ovals. His work was largely based on the work of his mentor, Danish astronomer Tycho Brahe. Kepler was able to use Brahe's precise measurements (made before telescopes) to determine, mostly by trial and error, three laws that described the motion of the then known five planets. Kepler published his first and second laws in 1609 in a book entitled *New Astronomy*. Ten years later, in 1619, Kepler discovered and published his third law.

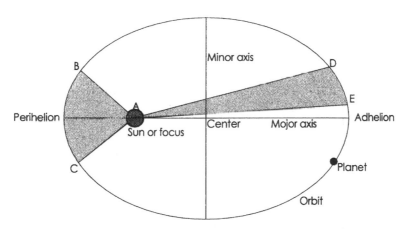

Explanation of Kepler's second law of planetary motion. Triangles ABC and ADE are of equal area. Distance BC is greater than DE. The planet travels distance BC at exactly the same duration as DE, meaning orbital speed is greater at perihelion than at aphelion. The difference between the major axis and the minor axis is referred to as eccentricity; if the focus coincided with the center, orbit would be a perfect circle and orbital speed would be constant and not fluctuate. Earth's eccentricity is very small indeed. Its minor axis is 98.5 percent of its major axis, thus almost a circle; therefore, increased velocity at perihelion is almost negligible.

- Kepler's first law, or the elliptical orbit law: Each planet moves in an elliptical orbit with the sun at one focus of the ellipse.
- Kepler's second law, or the equal-area law: A line from the sun to each planet sweeps out equal areas in equal time.
- Kepler's third law, or the law of periods: Orbits that are more distant have longer periods. The periods of the planets (T) are proportional to the 3/2 powers of the major axis lengths of their orbits (L). Expressed mathematically, then, this says that: T is proportional to L3/2 or T2 is proportional to L3. Although Kepler's laws were confirmed by observations and certainly seemed to be correct, he could not provide a theoretical explanation as to why they were correct. The theoretical explanation came about a century later in the form of Newton's law of gravity. Newton's universal gravitation can be used to prove that Kepler's laws do indeed precisely describe the motion of planetary bodies in orbit.

Eccentricity is an expression of the amount an ellipse is flattened; it is represented by a number between zero and one. Zero eccentricity means a circle is perfect, whereas an eccentricity of 1 means it is much flattened. Earth, for example, has an eccentricity of about 1.5 percent; Mercury has an eccentricity of about 20 percent.

Earth's aphelion equals 152,098,232 km, and its perihelion is 147,098,290 km, resulting in an eccentricity (the ratio between aphelion and perihelion) of 0.01671123. Earth orbits the sun precisely every 365.256363004 days, or 1.000017421 years.

Mercury is the innermost and smallest planet in our solar system, orbiting the sun once every 87.969 Earth days. Mercury's aphelion equals 69,816,900 km, and its perihelion is 46,001,200 km, giving Mercury's orbit the highest eccentricity (0.20563) of all planets in our solar system, and it also has the smallest axial tilt (2.11'). It completes three rotations about its axis for every two orbits, at an orbital speed of 47.87km/s, with an equatorial rotation velocity of only 10.892 kmh. As the smallest of all planets in our solar system, its diameter is only 4,879.4

km, just a little over one-third of the size of Earth. Mercury seems bright when viewed from Earth, with an apparent magnitude ranging from -2.3 to 5.7, but it is not easily observed, as its greatest angular separation from the sun is only 28.3°. Since Mercury is normally lost in the glare of the sun, Mercury can best be viewed in morning or evening twilight, unless there is a solar eclipse. It is an exceptionally dense planet because of the large relative size of its core. Surface temperatures range from about -183°C to 427°C, with the sub solar point being the hottest and the bottoms of craters near the poles being the coldest. Mercury's surface area is 7.48×10^7 km², or 0.147 times Earth's. Its volume is 6.083×10^{10} km³, equal to 0.056 Earths. Its mass is 3.3022×10^{20} tons or 0.055 times that of Earth.

Although all planetary orbits are elliptical, Venus's is almost circular, with an eccentricity of less than 0.01, rotating around its axis at 1.81 m/s, or 6.52 kmh, in the opposite direction to most planets in our solar system. Venus's perihelion is 107,476,259 km, and its aphelion is 108,942,109 km. It completes an orbit every 224.65 Earth days at an orbital speed of 35.02 km/s, or 126,072 kmh. The planet is named after Venus, the Roman goddess of love and beauty. After the moon, it is the brightest natural object in the night sky, reaching an apparent magnitude of -4.6, bright enough to cast shadows. Because Venus is an interior planet relative to Earth, it never appears to venture far from the sun; Venus reaches its maximum brightness shortly before sunrise or shortly after sunset, for which reason it is often called the Morning Star or the Evening Star. Venus's main diameter is 12,103.6 ± 1.0 km, or 0.949 x the Earth's, with surface temperatures reaching more than 460°C because of its close proximity to the sun.

With an apparent magnitude of 1.8 to -2.91, Mars is the fourth planet from the sun in our solar system. The planet is named after the Roman god of war, Mars. It is often called the Red Planet because the iron oxide on its surface gives the planet a reddish appearance. Mars's aphelion equals 249,209,300 km; its perihelion is 206,669,000 km. The eccentricity of Mars's orbit is 0.093315. The planet's orbital period equals 686.971 Earth days or 1.8808 years. Average orbital speed is

24.077 km/s. The equatorial radius of Mars is 3,396.2 ± 0.1 km, or 0.533 times that of Earth. Mars is one of the four terrestrial planets with a thin atmosphere, having surface features reminiscent of both the impact craters of the moon and the volcanoes, valleys, deserts, and polar ice caps of Earth. Mars's rotational period and seasonal cycles are likewise similar to those of Earth. Mars is the site of Olympus Mons, the highest known mountain in the solar system, and of Valles Marineris, the largest canyon. The smooth Borealis basin in the northern hemisphere covers 40 percent of the planet and may be a giant impact feature. Surface temperature ranges from -87°C to -5°C, with an average of -46°C. Until the first flyby of Mars occurred in 1965, by Mariner 4, astronomers speculated about the presence of liquid water on the planet's surface. This was based on observed periodic variations in light and dark patches, particularly in the polar latitudes, which appeared to be seas and continents; long, dark lines were interpreted by some as irrigation channels for liquid water. These straight-line features were later explained as optical illusions, yet of all the planets in the solar system other than Earth, Mars is the most likely to harbor liquid water, and thus to harbor life. Geological evidence gathered by unmanned missions suggest that Mars once had large-scale water coverage on its surface, while small geyser-like water flows may have occurred during the past decade. In 2005, radar data revealed the presence of large quantities of water ice at the poles and at midlatitudes. The Phoenix Lander directly sampled water ice in shallow Martian soil on July 31, 2008. Space probes operating at and on Mars in the past few years have completely revised our understanding of Mars, and in 2004, the first probe to sample particles from the sun returned to Earth. Mars has two moons, Phobos and Deimos, which are small and irregularly shaped. Mars is currently host to three functional orbiting spacecraft: Mars Odyssey, Mars Express, and the Mars Reconnaissance Orbiter. On the surface are the two Mars exploration rovers (Spirit and Opportunity) and several inert landers and rovers, both successful and unsuccessful. The Phoenix Lander completed its mission on the surface in 2008. Observations by NASA's now defunct Mars Global Surveyor show evidence that parts of the southern polar ice cap have been receding.

Mars can easily be observed from Earth with the naked eye. Its apparent magnitude reaches –2.91, a brightness surpassed only by Jupiter, Venus, the moon, and the sun. Mars has an average opposition distance (distance to Earth) of 78 million km but can come as close as 55.7 million km, which happened in 2003.

Jupiter is the fifth planet from the sun and the largest within our solar system. It is a gas giant with a mass slightly less than one-thousandth of the sun, but it has two and a half times the mass of all the other planets in our solar system combined. Jupiter is classified as a gas giant along with Saturn, Uranus, and Neptune. Jupiter's Aphelion equals 816,520,800 km, while its perihelion equals 740,573,600 km. The eccentricity of its orbit is 0.048775. Jupiter's orbital period is 4,331.572 Earth days, or 11.8592 Earth years, with an average orbital speed of 13.07 km/s. Jupiter's equatorial radius is 71,492 ± 4 km, or 11.209 times that of Earth's. Jupiter's equatorial rotation velocity is 12.6 km/s, or 45,300 kmh.

When viewed from Earth, Jupiter can reach an apparent magnitude of -1.6 to -2.94, making it the third-brightest object in the night sky after the moon and Venus, although Mars can briefly match Jupiter's brightness at certain points in its orbit. Jupiter's volume is equal to 1,321 Earths, yet the planet is only 318 times as massive. Likewise, Jupiter has a radius equal to 0.10 times the radius of the sun. In 1610, Galileo Galilei discovered the four largest moons of Jupiter (of a total of sixty-three)—Io, Europa, Ganymede, and Callisto (now known as the Galilean moons)—using a telescope; this is thought to be the first telescopic observation of moons other than Earth's. Galileo's was also the first discovery of a celestial motion not apparently centered on the Earth. It was a major point favoring Copernicus's heliocentric theory of the motions of the planets; Galileo's outspoken support of the Copernican theory placed him under the threat of the Inquisition.

Saturn is the sixth planet from the sun and the second-largest planet in the solar system, after Jupiter. Saturn is named after the Roman god Saturn. Saturn, along with Jupiter, Uranus, and Neptune, is classified as a gas giant. Together these four planets are sometimes referred to as the Jovian, meaning "Jupiter-like," planets. Saturn's aphelion is 1,513,325,783 km, while its perihelion is 1,353,572,956 km, and the eccentricity of its orbit is 0.055723219. Saturn has an average radius 9.4492 times larger than Earth's, at 60,268 ± 4 km. With an average orbital speed of 9.69 km/s, it takes Saturn 10,759 Earth days (or about 29.5 years) to finish one orbit around the sun. Saturn's equatorial rotation velocity equals 9.87 km/s or 35,500 kmh.

While Saturn has only one-eighth of the average density of Earth, owing to its larger volume, Saturn's mass is just over ninety-five times greater than Earth's. Wind speeds on Saturn can reach 1,800 kmh, significantly faster than those on Jupiter. Saturn is probably best known for its system of planetary rings, which makes it the most visually remarkable object in the solar system. Saturn has nine visible rings consisting mostly of ice particles and a small amount of rocky debris and dust; the rings are only about thirty meters thick. Sixty-two known moons orbit the planet, of which fifty-three are officially named, not including hundreds of moonlets within the rings. Titan, Saturn's largest and the solar system's second-largest moon (after Jupiter's Ganymede), is larger than the planet Mercury and is the only moon in the solar system to possess a significant atmosphere.

Saturn's rings require at least a 150 mm diameter telescope to resolve and thus were not known to exist until Galileo first discovered them in 1610. He originally interpreted the image as two moons, one on either side of the planet. It was not until Christiaan Huygens, using greater telescopic magnification, contradicted this concept and proved the "moons" to be a set of beautiful, bright rings. During one of Huygens's numerous observations, he also discovered Saturn's moon Titan. Saturn's atmosphere displays a striped pattern similar to Jupiter's, but Saturn's stripes are much fainter and are also much wider near the equator. About 80 km above the planet's surface lie ammonia ice clouds with temperatures of approximately -150°C.

With an apparent magnitude of 5.9 to 5.32, Uranus is the seventh planet from the sun, the third largest and fourth most massive in our solar system. Uranus's aphelion equals 3,004,419,704 km, and its perihelion 2,748,938,461 km. The eccentricity of Uranus's orbit is 0.044405586. It speeds through the cosmos at an average orbital speed of 6.81 km/s, or 24,516 kmh, completing one orbit every 30,799.095 Earth days, or 84.323 years. Uranus's diameter is 51,118 ± 4 km, or 4.007 times that of Earth. Its mass equals 8.681 × 10^{22} tons, or 14.536 times that of Earth, while it rotates around its axis at 2.59 km/s, or 9,320 kmh.

Although it is visible with the naked eye like the other five classical planets, it was not recognized as a planet by ancient observers, because of its dimness and very slow orbital speed. Sir William Herschel discovered Uranus on March 13, 1781, thereby expanding the then known boundaries of our solar system for the first time in modern history. It was also the first planet to be discovered with a telescope. The composition of Uranus, mainly ice and rock, is very similar to that of Neptune and is totally different from the compositions of the larger gas giants, Jupiter and Saturn, placing them in a separate category called the "ice giants." Uranus has the coldest planetary atmosphere in the solar system, with a minimum temperature of -224°C. Like the other giant planets, Uranus also has a ring system. It additionally has a magnetosphere and twenty-seven moons. An interesting feature about Uranus is that its north and south poles are located where most other planets have their equators. As seen from Earth, Uranus's rings sometimes appear to circle the planet perpendicularly, although in 2007 and 2008 these rings appeared edge-on, like a thin line. In 1986, images taken by space probe Voyager 2 showed Uranus as a virtually featureless planet, without the cloud bands or storms found on the other giants. However, terrestrial observers have noticed signs of seasonal change and increased weather activity in recent years as Uranus has approached its equinox (when day and night are of equal length). The wind speeds on Uranus can reach 900 kmh.

With an apparent magnitude of 8.0 to 7.78, Neptune is the eighth and farthest planet from the sun in our solar system. Neptune's aphelion equals 4,553,946,490 km, while its perihelion equals 4,452,940,833 km. Its orbital period lasts 60,190 Earth days, or 164.79 Earth years.

Named after the Roman god of the sea, it is the fourth-largest planet, with a diameter of 49,529 km, 3.883 times Earth's, and it is the third-largest planet by mass (1.0243×10^{23} tons, 17.147 times that of Earth). Equatorial rotation velocity equals 2.68 km/s, or 9,660 kmh. Neptune's orbital speed equals 19,548 kmh. Neptune was discovered on September 23, 1846, and was the first planet to be discovered by mathematical calculations rather than by telescopic observation. Its largest moon, Triton, was discovered shortly thereafter; the other twelve moons were discovered by using powerful telescopes, but not until the latter part of the twentieth century. Prior to the year 1846, Neptune was unknown to Earth's astronomers. It was just too far away to be seen with the naked eye. In the early 1800s, two scientists—John Adams in Great Britain and Jean Leverrier in France—each working independently, noticed strange behavior in the orbital movements of the planet. They concluded that perhaps the gravitational pull of some unknown planet was affecting Uranus. Working strictly on paper with mathematical calculations, each man independently concluded that another planet must exist.

Neptune has been visited only once by a spacecraft, Voyager 2, which flew by the planet on August 25, 1989. Neptune's atmosphere, similar to Jupiter's and Saturn's is composed primarily of hydrogen and helium, along with traces of hydrocarbons and possibly nitrogen; it contains a higher proportion of ices, such as water, ammonia and methane, causing the planet's soft blue appearance. In contrast to the relatively featureless atmosphere of Uranus, Neptune's atmosphere is notable for its active and visible weather patterns. For example, when space probe Voyager 2 flew by the planet in 1989, the Great Dark Spot was discovered in the southern hemisphere, which could be compared to the famous Great Red Spot on Jupiter. These weather patterns are driven by the strongest winds of any planet in the solar system, with wind speeds estimated as high as 2,100 kmh. Because of its great distance from the sun, Neptune's outer atmosphere is one of the coldest places in the entire solar system, with temperatures approaching -218 °C. Neptune, like all other outer planets, has a faint and fragmented ring system, which may have been detected during the 1960s but was indisputably confirmed in 1989 by Voyager 2.

Some time ago I was searching a NASA website for solar images, and I happened to come across an image called "Pale Blue Dot." After spending some time looking at the image and reading the caption, I gradually leaned back in my chair with awe and quietly thought, *Oh my God!* Never, ever before in my life did I realize how small and petite we as humans really are, and how majestic and great our Almighty God really is. Here I was looking at an actual, somewhat grainy, photograph of our planet Earth, hardly noticeable, taken by space probe Voyager 1 somewhere between February 14, 1990, and June 6, 1990. Earth showed up as a little pale blue dot from a distance of approximately 6.1 billion km away, almost forty-one times the distance from Earth to the sun. In this incredible photograph, Earth appears as a tiny speck of dust within the darkness of deep space in the middle of absolutely nowhere. It is indeed hard to imagine that Almighty God selected this particular tiny little pale blue speck of dust in such a vast universe as a habitat, a place we call our home, for almost 7 billion of us human beings.

On September 5, 1977, NASA launched Voyager 1. after completing its primary mission of photographing the planets of our solar system and leaving our solar system, it traveled through the emptiness of intergalactic space for a period of almost thirteen years at a speed of 40,000 kmh. Astronomer Carl Sagan (1934–1996) requested NASA to command Voyager 1 to turn its camera to the direction of our Earth and take a series of photographs across a great expanse of space, and it returned the image of the "Pale Blue Dot." Subsequently, Carl Sagan used the title of the photograph as the title of his 1994 book *Pale Blue Dot, A Vision of the Human Future in Space.* In a 2001 article by Space. com, Ray Villard and JPL's Jurrie Van der Woude selected this particular photograph as one of the top ten space science images of all time. Carl Sagan effectively reminded us that "all of human history has happened on that tiny pixel, which is our only home" In his book, Carl Sagan related his thoughts on a deeper meaning of the photograph with these memorable words:

From this distant vantage point, the Earth might not seem of particular interest. But for us, it's different. Consider again that dot. That's here, that's home, that's us. On it everyone you love, everyone you know, everyone you ever heard of, every human being who ever was, lived out their lives. The aggregate of our joy and suffering, thousands of confident religions, ideologies, and economic doctrines, every hunter and forager, every hero and coward, every creator and destroyer of civilization, every king and peasant, every young couple in love, every mother and father, hopeful child, inventor and explorer, every teacher of morals, every corrupt politician, every "superstar," every "supreme leader," every saint and sinner in the history of our species lived there— on a mote of dust suspended in a sunbeam. The Earth is a very small stage in a vast cosmic arena. Think of the rivers of blood spilled by all those generals and emperors so that, in glory and triumph, they could become the momentary masters of a fraction of a dot. Think of the endless cruelties visited by the inhabitants of one corner of this pixel on the scarcely distinguishable inhabitants of some other corner, how frequent their misunderstandings, how eager they are to kill one another, how fervent their hatreds. Our posturing, our imagined self-importance, the delusion that we have some privileged position in the universe, are challenged by this point of pale light. Our planet is a lonely speck in the great enveloping cosmic dark. In our obscurity, in all this vastness, there is no hint that help will come from elsewhere to save us from ourselves. The Earth is the only world known so far to harbor life. There is nowhere else, at least in the near future, to which our species could migrate. Visit, yes. Settle, not yet. Like it or not, for the moment the Earth is where we make our stand. It has been said that astronomy is a humbling and character-building experience. There is perhaps no better demonstration of the folly of human conceits than this distant image of our tiny world. To me, it underscores our responsibility to deal more kindly with one another, and to preserve and cherish the pale blue dot, the only home we've ever known.

It is indeed hard to imagine that this tiny, insignificant dot in this vast universe is the place where humanity had its origins and has existed for the last six thousand years.

It was not until about only 110 years ago that mankind invented and created a mechanical device that allowed him to leave the surface of Earth for the first time. Within sixty years of that historic moment, technology rapidly advanced to the point where mankind was able to leave Earth altogether and venture of into outer space.

Neil Alden Armstrong (b. August 5, 1930) was a test pilot, astronaut, and mission commander of the Apollo 11 lunar landing mission. On July 20, 1969, he and his fellow astronaut Buzz Aldrin were the first humans ever to set foot on the surface of our moon, and Armstrong was the first man ever to leave his footprint upon the surface of another celestial body, some 320,000 km from our Earth. Neil Armstrong commented: "It suddenly struck me that that tiny pea, pretty and blue, was the Earth. I put up my thumb and shut one eye, and my thumb blotted out the planet Earth. I didn't feel like a giant. I felt very, very small."

What is the anthropic principle? The anthropic principle is the law of existence of humanity. It is a well-understood fact that the existence of humanity depends on a large number of cosmological parameters whose numerical values must fit between an extremely narrow ranges of values. If any of these many variables were even slightly off, humanity could not exist. The extreme unlikelihood that all these variables line up in such perfect order and harmony has led some scientists and astronomers to propose that it was indeed Almighty God that designed, engineered, and created our universe, our galaxy, our solar system, and Earth. This is the anthropic principle: the universe, galaxies, solar system, and Earth appear to have been extremely fine-tuned to allow for humanity's existence. Some of these critical parameters include our solar system's location in the galaxy and Earth's place in the solar system. Our solar system is located in the outskirts, a quiet region of the Milky Way between two of the spiral arms. If our solar system were located much closer to the galaxy's center or to any of the spiral arms, our planet would be devastated by atomic radiation. If Earth were much farther away from the sun, our planet's water would freeze, and if we were much closer to the sun, our planet's water would boil. Water is one of the most vital elements for the existence of life.

"Water is actually one of the strangest substances known to science. This may seem a rather odd thing to say about a substance as familiar, but it surely true. Its specific heat, its surface tension, and most of its other physical properties have values anomalously higher or lower than those of any other known material. The fact that its solid phase is less dense than its liquid phase [ice floats] is virtually a unique property. The fact that ice floats allows aquatic life to exist in cold temperature zones" (Barrow 1996). Other important properties of water include its solvency, its cohesiveness, its adhesiveness, and its thermal properties. Water is common in three phases: solid (ice), liquid (water), and gas (water vapor). Rain is liquid water in the form of droplets that have condensed from atmospheric water vapor and then precipitated—that is, become heavy enough to fall under gravity. Rain is a major component of the water cycle and is responsible for depositing most of the fresh water on Earth's surface. Rain provides for many types of ecosystems, as well as water for hydroelectric power plants and crop irrigation.

Earth's atmosphere is composed of the perfect balance of specific gasses for life to flourish on Earth. The atmosphere consists of 78 percent nitrogen, 21 percent oxygen, and a small amount of other gasses, including helium and water vapor. If there happened to be too much of one of these gasses and not enough of another, our planet would either suffer a runaway greenhouse effect or it would be devastated by cosmic radiation.

In his book *The Unrandom Universe*, Sigmund Brouwer wrote that the odds against this essential atmosphere together with the water cycle forming on Earth by random chance alone are approximately one in a hundred trillion trillion! Earth is located in the habitable zone, the perfect distance from the sun where water can exist in three different forms: solid, or ice; liquid, or water; or as gas, in the form of water vapor.

If Earth were located much farther away from the sun, it would freeze. If it were much closer, it would boil or burn.

Earth's reflective albedo is the total amount of light reflected off the planet versus the total amount of light absorbed. If Earth's albedo were much greater than it is now, we would experience runaway freezing. If it were much less, we would experience an uncontrollable greenhouse effect.

A much stronger magnetic field would result in severe electromagnetic storm devastation; however, if Earth's magnetic field were much weaker, our planet would be devastated by cosmic radiation.

If Earth did not rotate as it does, the half facing the sun would become so hot that vegetation or any sort of life could not possibly survive, while temperatures on the dark side would drop to the point where life would not be possible either.

The twenty-three-degree tilt of Earth's axis provides for seasonal variation, allowing a wide variety of crops and vegetation to flourish. It prevents the north and south poles from becoming too cold and aids in regulating temperatures on Earth. Scientists have estimated that without Earth's tilt, approximately half of the earth's surface would be inhabitable, and less than half would be suitable for cultivation of crops and vegetables.

These are just few of numerous examples of how privileged our tiny little blue spot in this vast universe really is. "For he looketh to the ends of the earth, and seeth under the whole heaven; to make a weight for the wind: Yea, he meteth out the waters by measure. When he made a decree for the rain, and a way for the lightning of the thunder" (Job 28:24–26).

Gravity-time

Gravity and time are entities that have baffled the minds of the greatest thinkers of all times, including Galileo, Newton, and Einstein—three individuals respected and regarded as the most brilliant scientist the world has ever known.

The search for a scientific law that would adequately explain gravitation, or the force of gravity, began more than two thousand years ago when Greek philosopher Aristotle (384–322 BC) suggested that heavy objects would fall to the ground at a faster speed than lighter objects. For instance, a ten-pound object would fall to the ground at a speed ten times faster than a one-pound object. Although much progress has been made during the last two and a half millennia, scientists still have much to learn about the true nature of gravity.

Gravity is the mysterious force that holds and binds the entire universe together. There exists a gravitational force between all objects in the universe, and therefore it is referred to as universal gravitation. Without this universal gravitation, it would simply not be possible for the universe to exist. Gravity pulls you, as a person, to the center of Earth because there is gravitational force between you and Earth. There is also a gravitational force between you and the sun, as well as between you and the person sitting next to you. The reason you don't fly off to the sun is because the sun is too far away for its gravitational force to be effective. In simple terms, the sun is just too far away. The reason you do not bang into the person next to you is because people are much less massive than Earth. The gravitational force is proportional to mass, and it weakens when objects get farther apart from each other. For example, the moon is one-fourth the size of Earth, so the moon's gravity is much less than Earth's gravity—83.3 percent (or 5/6) less, to be exact.

An object is a thing that has physical existence; in other words, an object is anything made of matter. The total amount of an object's matter is called its mass. Gravitation is a force of attraction between objects in relation to their mass; in fact, the force of gravitational attraction is the definition of "weight." Applying the laws of universal gravitation, scientists can explain all motions we experience around us here on Earth as well as all motions of all heavenly bodies, such as planets, stars, and galaxies.

It was not until close to two thousand years after Aristotle that Italian scientist Galileo became interested in the nature of gravitational forces and started conducting practical experiments, building the foundation for today's understanding of how gravity works. In his famous experiment of dropping cannonballs of different weights from the fifty-four-meter Leaning Tower of Pisa, and later with careful measurements of balls rolling down inclines, Galileo showed that gravitation accelerates all objects at the same rate. A ten-pound and a one-pound cannonball, if dropped from the same elevation at the same time, reach the ground simultaneously, thus proving Aristotle wrong. Galileo also proved that substances of different specific gravities (also called relative density) descend at the same rate. For instance, 1,000 cm³ (10 × 10 × 10 cm), or 1 L, of water at its highest density (at 4° C) weighs 1 kg, while a block of solid gold of the same volume (1,000 cm³), weighing 19.3 kg, falls to Earth at the same velocity. He also correctly postulated that air resistance and buoyancy were the reasons lighter objects fell more slowly in an atmosphere. This was confirmed by astronaut David Scott during the 1971 Apollo 15 mission to the moon. Because the moon has no atmosphere, air resistance, or drag, does not exist on the moon. During the mission, astronaut Scott simultaneously released a hammer and a feather from the same elevation above the surface of the moon, and both objects fell at the same rate and reached the surface of the moon at the exact same time, although they fell at a much slower speed than they would have on Earth. (On the moon, gravitational acceleration is much less than on Earth, at approximately 1.6 m/s².) Close to four hundred years after Galileo's discovery, astronaut Scott demonstrated that in the absence of air resistance, all objects experience the same acceleration due to gravity. Galileo's work set the stage for the formulation of Isaac Newton's theory of gravity about eighty years later. He suggested that the strength of a gravitational field is numerically equal to the acceleration of objects under its influence, and its value at Earth's surface, denoted as g, is expressed as the standard average g=9.81 m/s²(=32.2 ft./s²). What this formula explains is that, ignoring air resistance, an object falling freely near the surface of Earth, increases its velocity by 9.81 m/s (22 mph) for every second of descent. Consequently, an object starting from rest will attain a velocity of 9.81 m/s after one second, 19.62 m/s after two seconds, and so forth, adding 9.81 m to each second of descent.

A free-falling object in an atmosphere, however, reaches its terminal velocity, or maximum speed, when the force of gravity equals the force of drag ($F_G=F_d$), causing the net force of the object to be zero, resulting in zero acceleration. For a person in free fall—a skydiver, for instance—the maximum speed he will reach is about 60 m/s, or 200 kmh. When he pulls the ripcord, his canopy opens, causing an increase in drag or air resistance and reducing his terminal speed to around 20 kmh, accommodating a safe and controlled landing.

Isaac Newton (1642–1727) is regarded by most modern scientists, as well as nonprofessionals, as one of the most brilliant scientists and mathematicians of all times. Amazingly, he was not an exceptionally brilliant student, but he was always extremely curious about how nature worked. Young Isaac's uncle became aware of his potential and persuaded his mother that he should be sent to Cambridge University to pursue his studies. During his early years at Cambridge, Newton came to realize that it is essential for scientists to rely on practical experimentation in order to understand the nature of the universe, rather than simply trusting what they can observe with their senses. In the summer of 1665, Newton left Cambridge in order to avoid the plague that was devastating Europe. During his absence from Cambridge, which lasted until the spring of 1667, he stayed at his mother's home in the hamlet of Wools Thorpe, in the county of Lincolnshire, and this is where he enjoyed some of the most productive months of his life. Not only did he lay the foundation for calculus, but he also began formulating the laws of motion and his theory of universal gravitation. According to the famous story, while walking in the orchard at the neighbor's house, an apple fell from a tree and landed on his head. However, Newton never admitted that the apple actually landed on his head; he just said that he saw an apple falling from a tree. However, this simple observation led him to speculate that the same gravitational forces that affect the motions of bodies on Earth also affect the motions of the moon and other celestial bodies. Newton wrote, "I thereby compared the force requisite to keep the moon in her orb with the gravitational force at the surface of the earth and found them to agree pretty nearly."

Newton's law of gravitation led directly to mathematical explanations of Galileo's falling object experiments and Kepler's laws concerning the motions of the planets. He formulated the following mathematical equation to explain his law of universal gravitation, which became just as famous as Einstein's energy equivalency equation, $E=mc^2$: $F= (Gm_1m_2)/d^2$. For the technically minded, G is the gravitational constant, which was measured by Henry Cavendish in 1798. ($G=6.672 \times 10^{-11}$ Nm^2/kg^2.) F is the force of gravity expressed in newtons, m_1 and m_2 are the weights of the masses of the two objects (Earth and an apple, or Earth and the moon, for instance), and d is the distance between them, measured from the centers of both masses. We notice with d^2 that gravity follows the famous inverse square law followed by many physical phenomena. Both Isaac Newton and Albert Einstein have dominated the field of gravity for the last 350 years. Einstein's theory, of course, represents the more recent and accurate conception of how gravity works.

Albert Einstein (1879–1955) was a German-born scientist who developed the general theory of relativity, or GR theory. For this achievement, he is often regarded as the father of modern physics and the most influential scientist of the twentieth century.

Einstein originally believed that the universe was static; that it had no beginning and had existed forever. He later admitted that this was the biggest blunder of his life! After he abandoned the idea that the universe was static and realized that it indeed had an origin, he also realized that the beginning of the universe was also the precise moment time had its beginning, and that the two therefore must be regarded as one, the same structure. Einstein theorized that time is a fourth dimension of space and that everything moves through it. He linked time and space together as one in this distinctive way, like a fabric, and he called this entity space-time. Einstein's radical way of thinking opened scientific minds to a completely new way of looking at the universe. Space-time affects everything in the universe.

In November of 1907, Einstein got a burst of inspiration that he described as "the happiest thought of [his] life." While working at the patent office in Berne, Switzerland, a thought suddenly occurred to him: *A person in free-fall will not feel his own weight.* He also concluded that without external clues, it is not possible to tell if you are being pulled down by gravity or accelerating upward, for instance in an elevator or rocket ship. Your legs will feel the same pressure and sensation; a ball will fall precisely in the same manner. The conclusion that gravity and acceleration are actually the same phenomenon resulted in the birth of his general theory of relativity.

Anyone having problems understanding exactly what Einstein's theory of general relativity entails should not feel bad, because there are actually very few people who do. In fact, when he first publicized his GR theory in 1915, the consensus was that only a handful of people in the entire world would be able to comprehend its meaning. However, Einstein's GR is not really that difficult to understand; it is merely a simple explanation of how things relate to each other, especially how they relate to an observer. For example, if you are driving an automobile on the highway at a speed of a 100 kmh and there is another automobile driving ahead of you, also at a 100 kmh, the speed of the other automobile relative to you, the observer, would be 0 kmh. However, if another automobile approaches from the opposite direction, also at a speed of 100 kmh, the relative speed of that automobile to you, the observer, would be 200 kmh. If an automobile approaches you from behind at a speed of 120 kmh and passes you, that automobile's speed relative to you, as an observer, would be 20 kmh, but for someone standing motionless on the sidewalk, the relative speed of that automobile would be 120 kmh. Your relative speed to the observer standing on the sidewalk would be 100 kmh. Similarly, if a pilot in the cockpit of an airplane turns on the cockpit light, to him the light is stationary and not moving, but for an observer on the ground watching the airplane fly by, the light moves at the speed of the airplane. Einstein's GR theory does seem to be simple indeed; however, at the astronomical level things seem to become more complicated, and a lot more difficult to understand.

Einstein was the first (and so far the only) scientist to offer an alternative concept of replacing Newton's universal law of gravitation, by declaring that gravity is not a force but a result of a distortion, or warpage, in the four-dimensional space-time continuum. Gravitational fields of heavenly bodies like stars, planets, and galaxies warp and distort the fabric of space-time. The more massive the object, the more space-time bends or distorts. "Newton, forgive me," he wrote. "You found the only way which in your age was just barely possible for a man with the highest powers of thought and creativity. The concepts which you created are guiding our thinking in physics even today ..."

In his book *The Grip of Gravity*, British scientist Prabhakar Gondhalekar explains, "For example, an apple falling to the Earth is no longer considered to be attracted by some mysterious force acting at a distance through space but instead rolls into a Space-time 'well' created by the gravity of the Earth." The gravity well, or curvature, is determined by the distribution of matter; the greater the density of matter in a given area, the greater the curvature of space-time in that area. A simple way to explain a gravity well is to picture a thin sheet of rubber, instead of the usual tarp, stretched across a trampoline frame. If a heavy object, for instance a bowling ball, is placed on the rubber sheet, the area around it is stretched or distorted. The amount of distortion depends on the mass of the object—the weight of the bowling ball, in this case.

Gondhalekar continues: "The Sun, being the most massive object in the solar system, causes the largest distortion of Space-time in its immediate vicinity. That curvature curves space further out, and so on.

The Planets are trapped in this well surrounding the sun." Not only did Einstein's theory of GR claim to bend, warp, and distort space-time, but it also suggested that distance and time are not absolute, as Newton had alleged before. He theorized that because of gravity, clocks would run at different speeds, ticking more slowly close to a gravitational mass like Earth or the sun, and fast-moving clocks would run slower than stationary clocks. Angles of a triangle would no longer add up to 180 degrees, and light passing close to a gravitational mass would actually bend, distort, and change direction. Both Newton's and Einstein's models suggest that light travels in a perfectly straight line when in the absence of gravity.

Einstein's theory of bending light was confirmed by observations made by a British expedition led by famous astronomer Sir Arthur Eddington during a solar eclipse in Africa on May 29, 1919. Eddington and his team conducted an experiment to confirm and photograph the precise locations of stars surrounding the sun during an eclipse. (This is the only time stars can be seen immediately adjacent to the sun.) To the shock of astronomers and scientists everywhere, Eddington's results confirmed what Einstein had predicted four years earlier—that the light from the stars would distort, or bend, as it passed by the sun. The photograph did not show the stars at their actual locations, but precisely where Einstein had predicted. International media reports of this event made Einstein instantly world famous and a celebrity, while Einstein remained somewhat unimpressed. Einstein's theory of bending light was also confirmed in 1979 with the discovery of two quasars, QSO 0957+561, only six seconds of arc apart. The two quasars were found to have identical redshifts and spectra. The probability for this to happen by pure chance alone is extremely limited, and astronomers positively concluded that this pair of quasars was actually the same quasar. The quasar's light was strongly distorted by the intervening mass of a cluster of galaxies in its path, causing its image to appear double as we view it from Earth.

In 1956, German American scientist Friedward Winterberg, at twenty-six years of age, proposed the idea of testing Einstein's theory of GR that time is not absolute (that clocks will run slower as their distance from a gravitational field increases) by placing atomic clocks in artificial satellites, even though atomic clock technology was still in its infancy and artificial satellites did not yet exist. Not only did Einstein predict the increase in clock speed with decrease of gravity, but he also predicted that moving clocks run slower as they move faster. The faster the clock moves, the slower it ticks. Friedward Winterberg, or anyone else at that time, had absolutely no idea that his proposal had actually laid the foundation for GPS, the most expensive, sophisticated, and accurate navigation system ever.

Thirty-one GPS satellites circle Earth twice a day at an altitude of 20,200 km, and at this altitude, acceleration due to gravity is approximately 4 percent of the acceleration we feel on Earth. Again, Einstein's GR theory proves to be correct, because at this altitude clock speeds do increase by about 45,900 nanoseconds per day. (One nanosecond is one billionth of a second; one nanosecond is to one second what one second is to 31.7 years.) GPS satellites travel at a speed of 14,000 kmh relative to the surface of our planet in order to maintain their precise orbit. Once again, Einstein's GR theory proves to be correct, because at 14,000 kmh, clocks do indeed slow down by approximately 7,200 nanoseconds per day compared to stationary clocks on Earth. Remarkably, these two effects cancel each other out somewhat, but not completely. The result is that orbiting atomic clocks in space still run faster than stationary clocks on Earth by about 38,700 nanoseconds per day. GPS satellites transmit signal information to Earth, where GPS receivers capture this information and use triangulation to calculate the user's exact location. Essentially the receiver measures the duration between when the signal was transmitted and when it was received. The system is designed to correct for these errors by adjusting the satellite clocks to run slower than their corresponding reference on Earth before they are launched, but periodic corrections must be performed at regular intervals to synchronize these clocks with time on Earth. If left unattended, a minor offset of 38,700 nanoseconds per day would be enough to cause navigational errors accumulating to not a few centimeters or meters, but to between 10 and 12 km per day. Without the proper implementation of Einstein's GR theory, the navigational functions of GPS would fail within two minutes!

While all this relativity theory may seem wonderful to the nonscientist, for the scientist it is simply not that easy. Things are not over yet. Relativity theories are still being tested and questioned. Einstein's GR theory seems to be correct at the astronomical level, but it does not work in the entire spectrum. GR theory has nothing to say about how gravity works when things get very small. At the level of subatomic particles, GR theory falls completely apart and has nothing to say.

On July 21, 1969, Neil Armstrong and Buzz Aldrin made history by being the first human beings ever to visit and set foot on the surface of another terrestrial body. First Neil Armstrong and then Buzz Aldrin actually walked on the moon. At 10:56, Armstrong exited the lunar module and spoke his legendary words, "That's one small step for man, one giant leap for mankind," as he made his famous first step on the moon. For about two and a half hours, Neil and Buzz collected lunar samples and conducted experiments. They also took many photographs, including images of their footprints, boot-shaped depressions in the gray moon dust. However, their footprints were not the only thing they left behind on the lunar surface; they left something else behind as well— something of much greater significance. Surrounded by their footprints, about one hundred feet from the landing site of the Eagle, they placed a forty-six-by-forty-six centimeter panel called the Lunar Laser Ranging Retro Reflector Array. This panel, studded with one hundred specially designed mirrors pointing at Earth, was placed there about an hour before they concluded their famous moonwalk. University of Maryland physics professor Carroll Alley was the principal investigator of the Apollo Lunar Laser Ranging Program, and he still follows its progress today. "Using these mirrors," explains Alley, "we can 'ping' the moon with laser pulses and precisely measure the distance from the Earth to the moon. This is a unique way to learn about the moon's orbit and to test both Newton and Einstein's theories of gravity." Alley also explains, "A laser pulse shoots out of a telescope on Earth, crosses the Earth-moon distance, and hits the reflector. Because the mirrors are 'corner-cube reflectors,' they send the pulse right back to the exact location where it originated. Back on Earth, telescopes intercept the returning pulse, sometimes not more than just a single photon." The duration of the beam's travel time is measured with atomic timing devices, which can divide a second into close to 9.2 billion fractions, pinpointing the moon's distance with staggering precision—within a few centimeters. This is incredible for an object at a distance of almost 385,000 km away. Since 1969, researchers and scientists have engaged the reflector to trace the moon's orbit precisely, and they made other remarkable discoveries as well. For instance:

- The moon is spiraling away from Earth at a rate of 3.8 centimeters per year; this is caused by the Earth's ocean tides.
- Because of minor "slashing" motions, we know that the moon probably has a liquid core.
- The universal force of gravity is very stable. Newton's gravitational constant G has changed less than 1 part in 100 billion since the laser experiments began.

The Lunar Ranging Program also revealed that the actual location of the moon is approximately ten meters from where Newton predicted it would be, according to his equation. Does this mean that Newton was wrong and his law of universal gravitation was flawed? British professor Brian Cox bluntly declares that it was indeed. Personally, I would rather contend that Newton was exceptionally close—within one-third of one-millionth of 1 percent. No, Newton's law of gravitation still remains a very valuable first approximation, which is adequate for almost all purposes, and certainly sufficient to allow NASA to perform their crucial calculations to launch Neil and Buzz to the moon and return them safely to Earth. In addition, Newton's law proves to be valuable for comparison with and understanding Einstein's version of gravity, which is much too complex for most purposes. Scientists and astronomers have also used the laser results to put Einstein's general theory of relativity to the test, with positive results. Einstein's equations predicted the characteristics of the moon's orbit, and laser ranging is capable of measuring it. Nevertheless, Einstein's theories are still constantly being tested. Some scientists, including Carroll Alley, believe his general theory of relativity is flawed. However, if there is a flaw, it will likely be lunar laser ranging that will discover the imperfections.

Accurate timekeeping has become critical to the functioning of today's society and is coordinated at an international level. The scientific explanation of what time really is has a tendency of becoming very complicated and utterly confusing. The most popular short answer to the question "What is time?" is actually very simple: time is not an event or an object; it cannot be touched or seen, but is rather a method of how to express the measurement of duration, the period between events or the duration of an event itself. A simple definition states, "Time is what clocks measure," or "Time is what keeps everything from happening at once." Up until recently, the measurement of time was based on astronomical phenomena—the rotation of Earth around its axis, termed a day, as well as the Earth's orbit around the sun, termed a year. However, in earlier times, astronomical events were not as well understood as they are today. For instance, during biblical times a year was considered twelve equal periods of thirty days each and was largely based on the phases of the moon. Great advances in accurate timekeeping were made during the seventeenth century by Galileo and, especially, Dutch scientist Christiaan Huygens, with the invention of the pendulum-driven clock in 1656. The pendulum increased the accuracy of clocks enormously, changing the margin of error from about fifteen minutes to fifteen seconds per day.

The most accurate timekeeping devices in use today are atomic clocks, which are calculated to be accurate to within one second in thirty million years and are employed to calibrate other clocks and timekeeping devices around the world. Since 1967, the International System of Units (SI) has defined the second as the duration of 9,192,631,770 cycles of radiation corresponding to the transition between the two energy levels of the cesium-133 atom, meaning the atomic clock ticks almost 9.2 billion times per second compared to an ordinary clock ticking only once every second. Wow! The new SI constant of the atomic second replaced the old astronomical second, which was still based on the motions of Earth. The GPS, or Global Positioning System, in coordination with the Network Time Protocol, is in use today to synchronize timekeeping systems across the globe.

The scientific big bang, or the theological moment of Creation, is accepted by both scientists and theologians as the origin of the universe. This event depicts the very beginning of everything; it was the very first thing that ever happened—the very first event, so to speak. According to scientists, prior to the big bang there was absolutely nothing. There was no yesterday, no past; there was not even space itself, for that matter. It is impossible for the limited human mind to comprehend what "nothing" really means. Theologians, on the other hand, have a completely different perception about the origin of the universe. They believe that God existed prior to the universe and that God has existed forever. In their view God had no beginning or origin, and therefore God is outside of the time dimension as well as outside of the laws of nature. They believe that, according to Genesis 1, God created the universe "by His Word." In any case, both parties do agree that the universe had a very definite origin, either by the creation of God or from nothing without reason. It was at this precise moment that time also had its beginning; this is the exact moment the clock started to tick, so to speak. This was the very first happening, or event, from which duration could be measured.

Logically, it would reasonable to accept gravity as one of the best-understood concepts in science. However, just the opposite is true. In many ways, gravity remains a profound mystery. Gravity provides a stunning example of the limits of current scientific knowledge. The world's greatest thinkers, including Galileo, Newton, and Einstein, have all failed to find a reasonable explanation for why gravity works the way it does. Both Newton and Einstein obviously came extremely close to discovering how it works, but neither had the faintest clue of why things work the way they do. Newton warned against using his law of gravity to view the universe as a mere machine, like a clock. A devout believer in God, Newton saw God as the Master Creator whose existence could not be denied in the face of all creation. He commented, "Gravity explains the motion of the planets, but it cannot explain who set the planets in motion. God governs all things and knows all that is or can be done." Newton's first law of motion asserts that a stationary body will remain at rest until some force generates its movement. Movement demands a mover. Gravity describes a current process, but the law of gravity does not address what caused planets to initiate their well-coordinated movements. Newton admitted, "This most beautiful system of the sun, planets, and comets, could only proceed from the council and dominion of an intelligent being. This being governs all things, not as the soul of the world, but as Lord over all; and because of his dominion, he is wont to be called 'Lord God' or 'Universal Ruler.' The Supreme God is being eternal, infinite, and absolutely perfect."

Although Einstein never believed in a personal God or Jesus Christ as a personal savior, he did acknowledge Almighty God as the Supreme Designer, Creator, and Architect of the universe. He stated, "I want to know how God created this world. I am not interested in this or that phenomenon, in the spectrum of this or that element. I want to know his thoughts. The rest are details … Everyone who is seriously involved in the pursuit of science becomes convinced that a spirit is manifest in the laws of the universe, a spirit vastly superior to that of man, and one in the face of which we with our modest powers must feel humble"

Einstein's famous reaction to the uncertainty principle was, "God does not play dice with the universe." Another one of his famous quotes was, "Science without religion is lame, religion without science is blind."

Stephan Hawking, who won global recognition with his 1988 book *A Brief History of Time*, is renowned for his work on black holes, cosmology, and quantum gravity. In his book, his views on religion are much in line with Einstein's, also sort of acknowledging God as the Supreme Creator. However, in his book *The Grand Design*, published in 2010, Hawking deviated from his previous views on creation and sort of sided with Richard Dawkins, who is perhaps the most visible and outspoken atheist scientist of all times. Dawkins bluntly declares that there is no God and continually ridicules people who do believe in God. Stephan Hawking proposes a different model for gravity. He bluntly declares that God did not create the universe and that the big bang was an inevitable consequence of the law of physics. He proclaims that because there is gravity, there is no need for God; however, he does so without proposing an alternative explanation of how and why gravity works the way it does or how gravity originated in the first place.

It is hard to predict if the scientific community will ever discover how Almighty God did create and sustain gravity. Perhaps like creation itself, such particulars may very well remain a mystery forever. The Creator is clearly an intimate part of every physical detail, gravity included. Can gravity ever be scientifically explained? Likely not! Can God ever be scientifically explained? Likely not! "He stretcheth out the north over empty space, and hangeth the earth upon nothing" (Job 26:7).

End of Times

Over the past decades, there has been much debate and speculation about the end of the days. A great number of authors and Bible scholars have extensively written on the subject, encouraging believers that we are literally living in the last days and that the end is approaching fast. It seems like each and every one of us has a keen interest in what the future has in store for us. Every time a major disaster happens, newspaper headlines compare it to prophetic Bible scripture and people start wondering whether we are really facing the end of the age. Debate about the end of times has always been a reality. Even before the New Testament books of the Bible were completed, there was already speculation that the return of Christ had taken place. The Thessalonians panicked and questioned Paul when they heard rumors that the day of the Lord was at hand and they had obviously missed the Rapture.

Science has little to say about the end of times other than proposing speculative doomsday threats, such as an asteroid colliding with our planet, contagious diseases, a runaway greenhouse effect, and resource depletion. Many well-informed scientists agree that climate change is the beginning of a global catastrophe; however, they are still debating whether climate change will result in the actual extinction of humankind or merely the end of civilization. Human activity has indeed made a significant impact on the surface of Earth. The increase in carbon dioxide in the atmosphere has increased by close to 30 percent since the start of the Industrial Revolution. There are multiple scenarios that can cause a devastating impact on Earth. Some of these scenarios may be brought about by humanity itself—for example, nuclear holocaust, misuse of nanotechnology, genetically engineered deceases, biological warfare, and catastrophe caused by physical experimentation. The end of times is one area where science and theology do not contradict each other; they both agree that humanity is rapidly approaching its grand finale.

The book of Revelation provides an accurate roadmap that precisely describes the future of humanity as well as the future of Earth. The book provides us with an abundance of passages that clearly reveal what is going to take place, and it presents many revelations about our imminent future. About one-quarter of the Bible is dedicated to the revelation of God and what His intentions are as biblical history approaches its appointed climax. In the midst of our out-of-control generation, it reveals a predetermined design that's unfolding exactly according to the Holy Scriptures. There are clear indications that the Rapture and the return of Christ may happen at any moment, and there are signs that will warn us of coming events.

The book of Revelation precisely describes the end times. It speaks of a one-world government; a godless, diabolic dictator; total regulation of all business; plagues; corruption; crime; rebellion; and devastating disasters. It seems to be an accurate description of what is happening today, but it was written more than two thousand years ago. Dr. Linus Pauling, winner of Nobel Prizes in 1954 and 1963, said he believes the greatest catastrophe in the world is approaching. This looming devastation might well result in a world war, which could destroy civilization and signify the end of humanity. Or it might take the form of mass starvation among a world population that has been doubling about every thirty-five years during recent decades. Civilization could very well come to a finale because of the collapse of the very systems on which it depends.

Prophetic scripture precisely prophesized the rebirth of the state of Israel; the continuous wars and threats of wars in the Middle East; a continuance of natural catastrophes such as earthquakes, tsunamis, and hurricanes; and the revival of Satanism and witchcraft. These events were predicted by the many prophets in the Old and New Testaments, including Jesus Christ Himself. They predict the coming of an antichrist, devastating wars that will bring humanity to the brink of destruction, and the incredible liberation of our desperate, dying planet, Earth. Only biblical prophecy can give us a clear insight of what truly and precisely will take place in the end times. Bible prophecy is the most fascinating and rewarding topic to study in the entire Bible.

Speculation about the exact time of the Rapture and the return of our Lord has always been a reality; many preachers and scholars have predicted dates, but these dates have always proven to be wrong for the very simple reason that there is absolutely no one who knows of that date except God the Father—not even the Son or the angels in heaven. Matthew 24:36 clearly states, "But of that day and hour knoweth no one, not even the angels of heaven, neither the Son, but the Father only." During the past centuries, there have been numerous influential people who have misled many people with false end-time predictions.

Individuals and religious groups who have dogmatically attempted to predict the day of the return of Christ, a practice referred to as "date setting," have been utterly embarrassed and discredited, as the predicted dates have come and gone without event. Some of these individuals and groups later offered excuses and revised or corrected target dates, while others simply published reinterpretations of the meaning of scripture to fit their current dilemma. Others explained that although the prediction appeared not to have come true, in reality it was completely accurate and had been fulfilled, though somewhat in a different way than was expected. On the contrary, many believe that the precise date of the return of Christ cannot be predicted or calculated, but they do recognize that the specific time frame that immediately precedes the return of our Lord can be identified. This time frame is often referred to as "the season." The primary section of scripture describing this opinion is Matthew 24:32–35; where Jesus is quoted teaching the parable of the fig tree, which many scholars believe to be the key that unlocks the understanding of the general timing of the Lord's return, along with the prophecies listed in the sections of scripture that precede and follow this parable.

This means that if there is any person or organization predicting the precise timing of the Rapture or the return of our Lord, as the Jehovah's

Witnesses are notorious for, it is probably wise to completely ignore these predictions, because they completely contradict Matthew 24:36 and other biblical end-time prophecies. We have always experienced that, after these predicted dates have passed, our world just kept on turning without significant change of any sort.

Over the last decade, there have been many scholars who have published extensive commentaries regarding the Rapture, the return of Christ, and the end of the days, including Hal Lindsey, Dr. Grant Jeffrey, John Hagee, Tim LaHaye, Jack Van Impe, Henry M Morris, Ed Hindson, and Chuck Smith, to name only a few. These writers offered incredibly accurate explanations of end-time events; however, several of these writers even went so far as to predict that the Lord would for sure, without any doubt, return at a specific time or date. Obviously, their predictions have failed, because of the simple fact that no one, other than Almighty God, can precisely, without failure, predict future events. Following are some samples of scholars who went beyond the prophecies of the Bible and thought they could calculate the exact date of the Rapture or the Second Coming of our Lord Jesus Christ.

Harold Lee "Hal" Lindsey (b. 1929) is a controversial American evangelist and author of several prophetic Christian books. Lindsey is a graduate of the Dallas Theological Seminary, where he studied with John F. Walvoord, author of the 1974 best-seller *Armageddon, Oil, and the Middle East Crisis*. In 1969 Lindsey authored his first book, *The Late Great Planet Earth*, published in 1970 by Zondervan of Grand Rapids, Michigan, shortly following the Six-Day War, which emphasized the popularity and support of ethnic Jews as the chosen people of God. *The Late Great Planet Earth* has been published in fifty-four languages, has reported sales of over thirty-five million copies, and is still in print today. Many of the books Lindsey later wrote are sequels, revisions, or extensions of his first book. The New York Times called *The Great Late Planet Earth* the "no. 1 non-fiction bestseller of the decade." For Christians as well as non-Christians of the 1970s, Lindsey's blockbuster served as a wake-up call to events soon to take place and already unfolding, all leading up to the greatest event of all times: the return of our Lord Jesus Christ. In 1981, Lindsey boldly declared that the Rapture would occur before December 31, 1981, based on biblical prophesies and astronomical events. He pointed out that the date of Jesus's promised return to Earth would happen a generation after the rebirth of the state of Israel. He also made references to the Jupiter effect, a planetary alignment that occurs every 179 years, which would supposedly lead to earthquakes and nuclear plant meltdowns. Chuck Smith also predicted that Jesus would probably return by 1981.

So what actually did happen on January 1, 1982? To be honest, not much of anything. All people, including born-again Christians, remained on Earth and continued their daily activities, and our world just kept on turning without significant change of any sort.

The year 1000 has gone down as the year causing the most hysteria over the speculated return of Christ. Society, rich and poor alike, for no justifiable reason, seemed to have a convincing notion that Jesus Christ would return on Jan 1, AD 1000. The magical number 1000 was likely primarily the sole reason for their expectation. During the month of December AD 999, everyone was on his or her best behavior; worldly goods were sold and donated to the poor, swarms of pilgrims headed east to personally meet the Lord Jesus at Jerusalem, commerce ceased functioning, businesses closed, agriculture was neglected, people quit their employment, and criminals were released from jails. It sure did look like the world had come to an end indeed. Then the year AD 999 turned into AD 1000. In spite of all their expectations, not much of anything happened. Our Lord was nowhere to be found—not in Jerusalem or anywhere else. Slowly but surely things returned to normal, and then our world just kept on turning without significant change of any sort.

In the year 2000 a somewhat similar incident occurred, although not related to the Rapture or the Second Coming of Christ, known as the Y2K problem, the millennium bug, the Y2K bug, or simply Y2K. During the 1960s to late 1980s it was common practice in computer software applications to express a year with two digits rather than four in order to save computer disk and memory space, because these resources were relatively expensive at that time. As the 1990s approached, experts began to theorize that this could possibly signify a major problem in application software. This particular bug existed not only in software applications but also in the firmware being used in the computer hardware. Generally speaking, there was a lot of speculation that this bug cold be a threat to and disrupt major industries, including utilities, banks, manufacturers, telecommunications firms, airlines, and air traffic control systems. Computer and system application companies introduced year-2000-compliant operating systems and system software. Companies around the world spent about $300 billion to go through their entire application source code to look for the Y2K bug and fix it. Almost everybody raced around to make themselves Y2K compliant before the fast-approaching deadline. Finally, when the big day came, many utilities and other companies switched off their main computers and put the backup computers in operation. When the clock changed to 12:01 on Jan 1, 2000, not a single problem was encountered. However, some questioned whether the absence of computer failures was the result of the preparations undertaken or whether the significance of the problem was grossly overstated. All the banks operated normally, no power outages were reported, airlines continued to operate without disruption, and our whole world just kept on turning without significant change of any sort.

William Miller (1782–1849) was an American Baptist preacher and the founder of an end-times movement that became so prominent it became worthy of its own designation: Millerism. Out of his direct spiritual followers grew several major Christian denominations, including Seventh-day Adventists. From his extensive studies of Bible prophecies, Miller determined that the Second Coming of Christ would take place, for sure, sometime between 1843 and 1844. A spectacular meteor shower in 1833 was interpreted as a divine sign from above and gave the movement unexpected forward momentum. The excitement and anticipation continued until March 21, 1844, when Miller's one-year time table came to a finale. However, the Second Coming never occurred, nothing out of the ordinary happened, and our world just kept on turning without significant change of any sort. After the failure of Miller's prediction, one of his colleagues, named Samuel Snow, claimed that the date should be corrected to October 22, 1844. That prediction also proved false and also failed, causing an abrupt finale for the Millerism movement, and again, our world just kept on turning without significant change of any sort.

Edgar C. Whisenant (1932–2001) was a former NASA engineer and devout Bible student. His book *88 Reasons the Rapture is in 1988* came out just a few months prior to when the Rapture was actually supposed to take place, between September 11 and September 13, 1988. Eventually, three hundred thousand copies of the book were mailed free of charge to ministers across America, and four and a half million copies were sold in bookstores as well as other places. Whisenant was quoted as saying, "Only if the Bible is in error am I wrong; and I say that to every preacher in town." As the final day approached, regular programming on the Christian Trinity Broadcast Network (TBN) was interrupted to provide special instructions on preparing for the Rapture. Whatever little time the book had, it certainly caused a lot of excitement and sensation in even some of the major Christian denominations. However, September 13 turned out to be just another day. The Rapture did not happen on this day or prior to it, and again our world just kept on turning without significant change of any sort.

However, after examining the 88 reasons the Rapture did not occur in 1988, Mr. Whisenant promptly published a new book called *89 Reasons the Rapture is in 1989*. This book was not taken very seriously, said to have sold only eighty-nine copies, and our world just kept on turning without significant change of any sort.

Charles Taze Russell (1852–1916), or Pastor Russell, was a prominent early twentieth century Christian Restorationist minister from Pittsburgh, Pennsylvania, and founder of what is now known as the Bible Student movement, from which Jehovah's Witnesses and numerous independent Bible Student groups emerged. After expressing a keen interest in the teachings of William Miller, Russell predicted that the return of our Lord would occur in 1914; however, our Lord probably decided to postpone His return to a more appropriate date, and again our world just kept on turning. Jehovah's Witnesses continued to predict end-time events to occur in 1918, 1920, 1925, 1941, 1975, and 1994. The year 1975 was likely selected because it was computed as the six thousandth anniversary of the creation of Adam in the garden of Eden in 4026 BC. The year 1994 was selected because Psalms 90:10 was interpreted as defining the length of a generation to be eighty years. Since 1914 plus 80 equals 1994, they predicted the return of Christ would occur in that year. However, our Lord did not return in 1975, or in1994, or on any of the other predicted dates, for that matter. What did happen? Absolutely nothing, and again our world just kept on turning without significant change of any sort.

Jack Van Impe Ministries is the sponsor of the largest Evangelical Christian television program devoted to end-time prophecy. On his website he discussed his book *On the Edge of Eternity*, in which Mr. Van Impe predicted that the year 2001 would "usher in international chaos such as we've never seen in our history." He predicted that in 2001, and the years following, the world would experience "drought, war, malaria, and hunger afflicting entire populations throughout the [African] continent ... By the year 2001, there would be global chaos." He also predicted that Islam would surpass Christianity, even though only 19 percent of the world's population follows Islam compared to 33 percent for Christianity. A one-world church would emerge, controlled by "demonic hosts." Temple rituals, including animal sacrifices, would resume in Israel.

Herbert W. Armstrong (1892–1986) was the founder of a religious organization called the Worldwide Church of God, and he published a book called *1975 in Prophecy!* The book warned of a catastrophic drought, enslavement of Americans and Britons, nuclear holocaust, and he the return of our Lord Jesus as a dictator. However, 1975 turned out to be just another year and did not see the arrival of this predicted "dictator," and again our world just kept on turning without significant change of any sort.

The idea that the world would end on December 21 2012 was a rather recent phenomenon. It originated from speculations relating to the ancient Mayans, an advanced civilization, which was once centered in what is now Mexico. The Mayans erected elaborate temples throughout their lands and religiously tracked the movements of the stars and planets. Through their observations, they developed the Mesoamerican calendar, a rather sophisticated method for keeping track of time. While a detailed explanation of this calendar is beyond the scope of this article, for simplicity's sake, we'll note that two parts of this calendar system, the Tzolk'in and the Haab, synchronized a period of fifty-two years. In order to track more than fifty-two years of time, the Mayans relied on the use of the Long Count calendar, which moved in 5,125-year cycles. According to many subscribers of the 2012 end-of-the-world theory, the Mayan Long Count calendar would end its 5,125-year cycle precisely on December 21, 2012.

There was much debate about what exactly was supposed to happen on that date. Many people thought it served as an indication that the world would definitely come to an end, with an overabundance of natural disasters like earthquakes, tsunamis, typhoons, solar storms, and a magnetic pole reversal. Some even suggested Earth would collide with an asteroid, a comet, or some other heavenly body. While Mayan scholars and archeologists vigorously disputed these claims, public fascination with the occasion continued. The publication of numerous books, such as New Age author Daniel Pinchbeck's *2012: The Return of Quetzalcoati* (2006) and mega-bestselling novelist Dan Brown's *The Lost Symbol* (2009) only aided in expanding the 2012 phenomenon. However, the Mayan Long Count calendar never did provide a glimpse into humanity's future or made any reference to end times. Obviously, nothing happened on December 21, 2012, and again, our world just kept on turning without significant change of any sort.

As a direct result of the ongoing, undeniable failure of the modern-day "prophetic end-time predictions," many believers have come to the conclusion that the study of Bible prophecy should be restricted to the study of the Bible itself, ignoring predictions made by so-called experts who neglect to pay any attention to Matthew 24:36: "But of that day and hour knoweth no one, not even the angels of heaven, neither the Son, but the Father only."

So what does the Bible really have to say about the end of the age, and when can we reasonably expect the Rapture and the Second Coming of Christ without calculating a precise date?

In 1961 Arthur E Bloomfield (1895–1980) published his book *The End of The Days*, in which he expanded on the prophecies of Daniel in a comprehensive text. Comparing the book of Daniel to the Olivet Discourse and the book of Revelation, Bloomfield revealed a precise sequence of events for the last days. However, he did so without predicting a specific date for the return of our Lord. The Olivet Discourse is the name given to the teaching by Christ on the Mount of Olives. The Olivet Discourse, the book of Daniel, and the book of Revelations are all interrelated and form a prophetic trilogy. In the Olivet Discourse, Jesus's teachings reach back to the book of Daniel and forward to the book of Revelation without unnecessary repetition, combining all three into one easy-to-understand unit to form a bridge that connects Daniel to Revelation. The Olivet Discourse is recorded in Matthew 24:1–25:46. Passages describing the same events are also found in the Gospels of Mark (13:1–37) and Luke (21:5–36). Christ's discourse is a response to questions asked by the disciples, based on what Jesus told them in Matthew 24:1–2, Mark 13:1–2, and Luke 21:5–6.

During His ministry here on Earth, Jesus revealed an abundance of details about what we should expect to indicate His Second Coming and the end of the age. Another detailed analysis, written by theologian Ray Stedman (1917–1992), refers to the same events in his book *Olivet Prophecy: The most detailed prediction in the Bible*. According to Stedman, "there are many predictive passages in both the Old and New Testaments, but none is clearer or more detailed than the message Jesus delivered from the Mount of Olives. This message was given during the turbulent events of the Lord's last week before the cross."

> And as he sat on the Mount of Olives, the disciples came unto him privately, saying, tell us, when shall these things be? And what *shall be* the sign of thy coming, and of the end of the world? (Matthew 24:3)
>
> For many shall come in my name, saying, I am the Christ; and shall lead many astray. (Matthew 24:5)

> And ye shall hear of wars and rumors of wars; see that ye be not troubled: for *these things* must come to pass; but the end is not yet. (Matthew 24:6)
>
> For nation shall rise against nation, and kingdom against kingdom; and there shall be famines and earthquakes in divers places. (Matthew 24:7)
>
> …but all these things are the beginning of travail. Then shall they deliver you up unto tribulation, and shall kill you: and ye shall be hated of all the nations for my name's sake. (Matthew 24:8–9)
>
> And many false prophets shall arise, and shall lead many astray. (Matthew 24:11)
>
> And this gospel of the kingdom shall be preached in the whole world for a testimony unto all the nations; and then shall the end come. (Matthew 24:14)

In the Olivet Discourse, Jesus sat down with Peter, James, John, and Andrew and explained the end-times episode. In it, He spoke of the abomination of desolation occurring, and then the last half of the seven-year period would begin with the Great Tribulation period.

> When therefore ye see the abomination of desolation, which was spoken of through Daniel the prophet, standing in the holy place (let him that readeth understand), then let them that are in Judaea flee unto the mountains. (Matthew 24:15–16)
>
> But when ye see the abomination of desolation standing where he ought not (let him that readeth understand), then let them that are in Judaea flee unto the mountains. (Mark 13:14)
>
> But when ye see Jerusalem compassed with armies, then know that her desolation is at hand. Then let them that are in Judaea flee unto the mountains; and let them that are in the midst of her depart out; and let not them that are in the country enter therein. (Luke 21:20–21)

The prophecies of the biblical end times can be compared to trying to solve a jigsaw puzzle; we know all the pieces are present, but we just need to find the right location for each of the pieces to provide us with the complete picture. The most difficult part is the beginning of the puzzle, because we need to start with one piece only, most likely a corner, and then we must search for the next piece out of all the remaining hundreds or perhaps thousands of pieces, each with a different size, shape, and color. The more pieces we succeed in finding the right location for, the easier it becomes to find the spot where the next piece fits. Toward completion, when there are only a few pieces left to select from, it becomes much easier to determine where the next piece will fit. Finally, when there is only one piece left, there is only one possible location, and the puzzle is solved.

Biblical prophecy works in much the same way. When we start our study, we are overwhelmed with the hundreds or perhaps thousands of prophetic verses and statements located throughout the entire Bible. We just need to sit down, study, and find the correct way to combine all these verses and predictions together in the right sequence. After examining world events that have already taken place, representing the part of the puzzle that is already completed, we reasonably will have some idea of what we can expect in the near future. In order to see the completion of the prophetic puzzle, however, we do not have to do participate; we just need to sit back, relax, and watch how the pieces fall into place. In biblical prophecy, there is nothing we can do to assist, because Almighty God is ultimately in control of the future and the destination of humanity and our world. He will put all the pieces of the prophetic puzzle in the right spot, unlike us trying to complete our puzzle.

According to the Olivet Discourse, Daniel, and the book of Revelation, here is a short general timeline of issues and events leading to the end of the age, giving us an idea what we can expect in the near future. This starts with Daniel's much-discussed seventy weeks, which began with the edict of Artaxerxes, a decree in where the Jews were allowed to return to Jerusalem to rebuild their city and temple after their exile in Babylon for seventy years. At the end of week sixty-nine, the

Messiah was "cut off," referring to the crucifixion of Christ, leaving the last week, or the seventieth week, reserved for the end-of-the-age events, referred to as the Tribulation. The fulfillment of Daniel's prophecy is of such magnitude that we saw an estimated 40 percent of the prophetic puzzle completed.

After the destruction of Jerusalem in AD 70, the "times of the Gentiles" began, and the Jews were dispersed among the "nations of the world." The times of the Gentiles began with the apostle Paul's missionary work in the first century. Prior to this time, the Word of God was preached mainly to the Jewish population. Many end-times prophesies are directed to the Jewish population living within Israel. Since the destruction of Jerusalem in AD 70, Israel had not been an independent Jewish state for nearly two thousand years. Therefore, against incredible odds, a major part of prophesy was fulfilled when the Jews did, in fact, return to reestablish the nation of Israel with its ancient language, Hebrew—their homeland, which God had promised to their forefathers, Abraham, Isaac, and Jacob. Incredibly, on May 15, 1948, David Ben Gurion declared the state of Israel to be a sovereign nation. Fulfillment of this prophecy is of incredible enormity; it forms the foundation for the beginning of the end of the age. This prophecy described the rebirth of the state of Israel "during the last generation." Here we saw another estimated 45 percent of the prophetic puzzle being completed.

The sequence of events we can expect after this incredible occurrence are very clear; we know exactly *what* is going to happen; we just don't have any idea *when* this is going to happen. The next major event will be the Rapture; this is where all born-again Christians, or saints, will be taken from the earth to meet the Lord in the air and be taken to heaven, according to Timothy. The stage is already prepared for this episode; this significant event could very well happen within the next few moments, or perhaps tomorrow, or next month, or maybe not for several years; nobody really does know, or has any logical means of predicting or calculating, the precise moment. Remember Matthew 24:14: "And this gospel of the kingdom shall be preached in the whole world for a testimony unto all the nations; and then shall the end come."

This verse tells us that the end will not come until the word of God has been preached to each and every soul on the face of the earth. Incredibly, modern technology has provided the means for this to actually happen today. There is hardly a castle, villa, home, cave, hut, or igloo today in which one does not find a flat-screen TV with global reception, and occasionally TV preachers do preach the Word of God rather than beg for money.

At the Rapture, the prophetic puzzle for born-again Christians will be 100 percent solved, because they will be taken from the earth and "Raptured" to heaven to be with the Lord. For them, earthly life has come to a finale, and they will spend eternity with God. They do not have to endure the horrors of the Tribulation, unlike the ones that are left behind. For the ones that are left behind, things will become very unpleasant; every event will be like another horrible surprise.

The Tribulation will follow shortly after the Rapture. The Tribulation is the final 2,520 day (seven-year) period before the Second Coming of Christ. This is the period when God will pour out His wrath on this corrupt, sinful, and disobedient world. This is the time when the Antichrist, or the Beast, powered and inspired by the number-one spiritual enemy, Satan, will rise to power and initiate a seven-year peace treaty with the newborn nation of Israel. However, after three and a half years he will dishonor the treaty, resulting in the gathering of the armies of the nations of the world in the valley of Megiddo and, consequently, the battle of Armageddon, during which the Lord Jesus will return to the earth and defeat the armies of the Antichrist, and Satan will be bound for a thousand years. Then the Lord will set up His glorious kingdom here on Earth, where He will reign over the chosen people and Christians on Earth for one thousand years, in a world of peace without deceptive Satan (Revelation 20:1–6, Isaiah 60–66). At the end of millennium, Satan will be unleashed and will try to deceive the world one more time, and he will cause war once more (Revelation 21:7–9). Then the world will end, and Christians will enjoy a new heaven and new earth, where everybody that made it to paradise—those whose names are written in the Book of Life—will be happy and smiling for eternity (Revelation 21). The wicked and nonbelievers will go to the everlasting lake of fire along with Satan (Revelation 20:10–15).

All these events represent the remaining few pieces of the prophetic puzzle; we can see their location, and we know exactly where they fit, but we must just wait for Almighty God to place them in the right place, because we have absolutely no idea precisely when they will occur.

The Rapture

The Rapture will indeed be one of the world's most spectacular events; it is speculated to occur in the near future. This is where literally millions of people will disappear or vanish from the face of the earth in an instant. This event will have such global significance that it will shake the entire population of the world. The Rapture is the instantaneous removal of God's people, or the saints, from the earth prior to the Great Tribulation. If Niagara Falls were to suddenly vanish from the face of the earth, national broadcasting stations like CBC, CTV, ABC, NBC, CBS, and CNN would instantly break from regular programming and provide live coverage of such a unique event. If multitudes of people were to disappear without a trace, in the blink of an eye—the media frenzy cannot be imagined. If only one car or one airplane crashed without the evidence of there having been a driver or pilot, it would be a major international news item; however, after the Rapture, the media will have innumerable similar events to report. Consider the countless crucial positions Christians hold in the workplace. Many businesses will be completely paralyzed by the loss of their key personnel; the economy will suffer a devastating blow. Millions of people who had friends and family members Raptured will be totally terrified. Mortgage payments, car payments, and credit card payments will no longer be made, causing financial institutions to sink and collapse. The world will literally be at the brink of disaster. The worldwide instability of the Rapture will create a media event that will surpass any news event from the past. When you turn on your flat-screen TV, you will find twenty-four-hour coverage on every channel. The leaders of the world will be calling emergency meeting after meeting, and churches will be filled to overflowing. Many Christians in the prophetic realm today believe that when the Rapture takes place, very few individuals that are left behind will understand the true reason for its occurrence. What will be the explanation for this sudden occurrence? Who will be left behind, and when can we expect this to happen?

Jesus repeatedly stated that His return for the church would be a surprise. The Lord even went beyond that by saying He would return "as a thief in the night" at a time when believers generally won't be expecting Him. If the church were required to go through the seven-year tribulation, the New Testament writers would have given us more warning to prepare for these terrifying end-times events. On the contrary, the New Testament writers repeatedly tell the church to be comforted by the "coming of the Lord" (1 Thessalonians 4:18). The word "comfort" alone strongly implies that the Rapture will take place before the Tribulation.

At the Rapture, all saints, or born-again Christians, will be gathered together in the air to meet Christ prior to his Second Coming. Apostle Paul stated in 1 Thessalonians 4:16–18, "For the Lord himself shall descend from heaven, with a shout, with the voice of the archangel, and with the trump of God: and the dead in Christ shall rise first; then we that are alive, that are left, shall together with them be caught up in the clouds, to meet the Lord in the air: and so shall we ever be with the Lord. Wherefore comfort one another with these words."

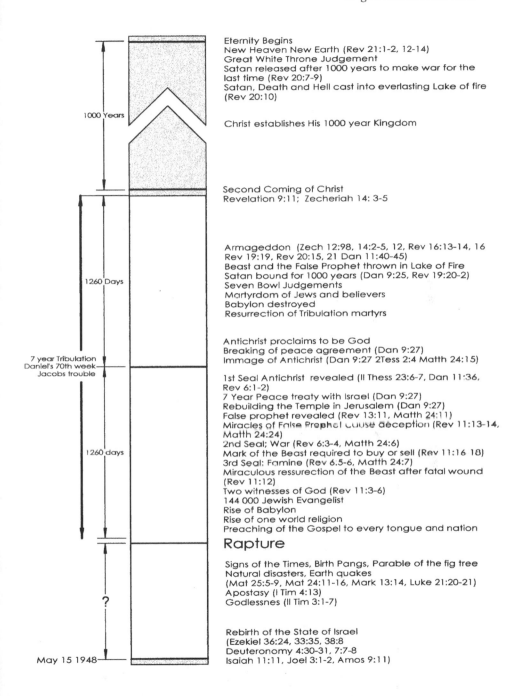

Eternity Begins
New Heaven New Earth (Rev 21:1-2, 12-14)
Great White Throne Judgement
Satan released after 1000 years to make war for the last time (Rev 20:7-9)
Satan, Death and Hell cast into everlasting Lake of fire (Rev 20:10)

Christ establishes His 1000 year Kingdom

Second Coming of Christ
Revelation 9:11; Zecheriah 14: 3-5

Armageddon (Zech 12:98, 14:2-5, 12, Rev 16:13-14, 16
Rev 19:19, Rev 20:15, 21 Dan 11:40-45)
Beast and the False Prophet thrown in Lake of Fire
Satan bound for 1000 years (Dan 9:25, Rev 19:20-2)
Seven Bowl Judgements
Martyrdom of Jews and believers
Babylon destroyed
Resurrection of Tribulation martyrs

Antichrist proclaims to be God
Breaking of peace agreement (Dan 9:27)
Immage of Antichrist (Dan 9:27 2Tess 2:4 Matth 24:15)

1st Seal Antichrist revealed (II Thess 23:6-7, Dan 11:36, Rev 6:1-2)
7 Year Peace treaty with Israel (Dan 9:27)
Rebuilding the Temple in Jerusalem (Dan 9:27)
False prophet revealed (Rev 13:11, Matth 24:11)
Miracles of False Prophet cause deception (Rev 11:13-14, Matth 24:24)
2nd Seal; War (Rev 6:3-4, Matth 24:6)
Mark of the Beast required to buy or sell (Rev 11:16 18)
3rd Seal: Famine (Rev 6.5-6, Matth 24:7)
Miraculous ressurection of the Beast after fatal wound (Rev 11:12)
Two witnesses of God (Rev 11:3-6)
144 000 Jewish Evangelist
Rise of Babylon
Rise of one world religion
Preaching of the Gospel to every tongue and nation

Rapture

Signs of the Times, Birth Pangs, Parable of the fig tree
Natural disasters, Earth quakes
(Mat 25:5-9, Mat 24:11-16, Mark 13:14, Luke 21:20-21)
Apostasy (I Tim 4:13)
Godlessnes (II Tim 3:1-7)

Rebirth of the State of Israel
(Ezekiel 36:24, 33:35, 38:8
Deuteronomy 4:30-31, 7:7-8
Isaiah 11:11, Joel 3:1-2, Amos 9:11)

Labels on timeline:
1000 Years
1260 Days
7 year Tribulation Daniel's 70th week Jacobs trouble
1260 days
?
May 15 1948

Graphical representation of the Time of the End

This means that people that have died as born-again Christians before us shall rise first. Then born-again Christians that remain will instantly be Raptured to the presence of the Lord without the sting of death. Further biblical verses relating to the Rapture can be found in Matthew 24:31, 40–41; 1 Corinthians 15:50–53; 1 Thessalonians 4:13–17; John 14:1–3; 2 Thessalonians 2:3, 6–7; 2:5, 9–10; Luke 21:34–36; Isaiah 26:19–21; and Revelation 4:1–2. All saints, or born-again Christians, will be raised up at the Rapture and changed in the twinkling of an eye. "Behold, I tell you a mystery: We all shall not sleep, but we shall all be changed, in a moment, in the twinkling of an eye, at the last trump: for the trumpet shall sound, and the dead shall be raised incorruptible, and we shall be changed. For this corruptible must put on incorruption, and this mortal must put on immortality" (1 Corinthians 15:51–53).

We will be taken by the Lord for the judgment seat of Christ and the Marriage Supper of the Lamb, and then we will return seven years later at the Second Coming. After the battle of Armageddon, the judgment of the nations, and the destruction of the wicked, we will enter into the Millennial Period, where Christ shall reign from Jerusalem as King over the whole earth. We will serve our Lord, and we will reign with Him. After the saints, or believers, are Raptured, there are many that will be left behind. Who will be left behind? Simply, all those who ignore and deny Almighty God and the ones that do believe in Him but never have accepted the salvation of Jesus Christ. In this day and age, many people actually believe and profess to be saved but are totally ignorant as to what it actually means to be a true born-again Christian. Today's world's understanding of a believer may be very different from what the Bible proclaims.

Nicodemus asked Jesus the all-important question, what it takes to enter the Kingdom of God. Jesus answered and said unto him, "Verily, verily, I say unto thee, except one be born anew, he cannot see the kingdom of God" (John 3:3). Matthew 7:21 clearly acknowledges, "Not everyone that saith unto me, Lord, Lord, shall enter into the kingdom of heaven; but he that doeth the will of my Father who is in heaven." Only those who are truly saved are the ones that will be Raptured.

The problem with today's church age is that it is selfish, boasting in their wealth and prosperity, not realizing that it is truly lost, and in desperate need of spiritual revival. "But know this, that in the last days grievous times shall come. For men shall be lovers of self, lovers of money, boastful, haughty, railers, disobedient to parents, unthankful, unholy, without natural affection, implacable, slanderers, without self-control, fierce, no lovers of good, traitors, headstrong, puffed up, lovers of pleasure rather than lovers of God; holding a form of godliness, but having denied the power therefore. From these also turn away." This is how 2 Timothy 3:1–5 describes mankind in the last days prior to the Rapture, and it represents an accurate portrayal of how mankind behaves today. Wickedness flourishes even in the largest professedly Christian churches.

The people left behind will be unaware of their sad and hopeless circumstance. Today's church have never considered themselves sinners and in need of a savior. Many church members today actually believe that they are saved because they think they are good people, do good deeds, and acknowledge God. They believe that church membership, some sort of Christian affiliation, or perhaps a sporadic prayer will get them to heaven. But if they never actually accepted God's Son as Lord and savior of their life, they are going to miss the boat. If they believe that their occasional or perhaps regular church attendance will provide them with the key to heaven and eternal life, they are grossly mistaken. I know many people whom I know in my heart are not saved, but they believe they are saved just because they periodically attend church services. These are the people that may have a vague idea of what the Rapture is all about, but they will be left behind and won't really understand why. Unfortunately, these are the people that will endure the horrors of the Tribulation Period. However, some of these people will eventually come to realize why they were left behind and to also realize that they still have a chance to turn to Christ and recover their positions. They vaguely remember God's Word and the biblical warnings about the Rapture, the end-times, and now they realize that these events are actually happening. People left behind and living during this period will have only one option for salvation, be saved by committing their lives to

Jesus Christ and by refusing to accept the mark of the Beast. There will be widespread genocide and persecution, especially of Christians and people who refuse to accept the mark of the Beast, as required by the Antichrist, the coming world leader, as discussed later in this chapter.

"And he causeth all, the small and the great, and the rich and the poor, and the free and the bond, that there be given them a mark on their right hand, or upon their forehead; and that no man should be able to buy or to sell, save he that hath the mark, *even* the name of the beast or the number of his name" (Revelation 13:16–17). Anyone that refuses the mark of the Beast will be considered an outcast and will be persecuted and martyred. However, it is far better to die for a heavenly Savior and see the kingdom of heaven than to live and submit to a satanic worldly leader who unjustly executes the innocent. In other words, accepting the mark of the Beast literally means selling your soul to the devil! People that are left behind and accept the mark of the Beast will never see heaven. For it is only through faith in the Lord Jesus Christ and a truly repentant heart that anyone can be saved.

Satan, our spiritual enemy, is just as familiar with end-time events as the saints or born-again Christians. He is already preparing the world for the Rapture and is already playing his cards by recruiting atheists like Richard Dawkins to spread rumors about aliens and warn about alien abductions to explain away and divert the true events about the Rapture. After the Rapture occurs, these are the people whose faces will be seen on television screens; their pictures will appear on the front pages of newspapers proclaiming "I Told You So." However, these people will only be instruments of Satan and an integral tool of Satan's deceit.

Although I have never personally encountered either a UFO or an alien, this does not necessarily mean, or prove, that they do not exist. There are indeed many credible sources, including doctors, lawyers, and even astronauts that claim UFOs and aliens do exist and that they have had encounters with them. One of the most famous UFO stories in history is that in 1947 an alien craft crashed in the New Mexico desert near Roswell and that civilians arriving at the scene witnessed dead and injured alien bodies. When the military arrived, they captured and transported the craft and the alien corpses, performed autopsies, and initiated a massive cover-up. However, when millions of people suddenly disappear from the face of Earth, I would rather attribute the incident to the biblical Rapture than to alien abductions. I do not believe that there will be alien abductions, nor will there be alien spaceships, for the very simple reason that aliens are not likely to exist on Earth; they are just too far away. In spite of all technical and scientific advances of our modern times, other life forms in the universe have yet to be discovered, and astronomers are convinced that it is not likely that there are other planets or heavenly bodies capable of supporting any sort of life that can possibly exist within commutable distance from Earth. Nowhere in the Bible is there evidence or any suggestion that other life forms do, or may, exist somewhere else in our universe, our even our own galaxy. However, if UFOs do exist, they will most likely be within our own unexplained realm rather than from some unknown planet or heavenly body quadrillions of miles away. Throughout history odd lights, strange apparitions, and mysterious creatures have appeared to mankind in the skies above and on Earth. God also came to Earth, revealed Himself to us, told us the truth, and lived among us as one of us. The Bible does teach that there are angels and demons and that there is warfare in the heavenly spheres. Demons are the fallen angels that once obeyed God but later rebelled with Satan and turned against God. Is it possible that the UFO phenomenon is actually a glimpse of a battle raging in the hidden spiritual realms?

Social, political, economic, and technological changes are occurring in all aspects of our "global society" today. The collapse of ethical and moral values throughout society and government in the twentieth century was an unparalleled change, and the twenty-first century is proving to be even more dramatic. At no time in history have more changes occurred in the scientific, technological, and political structure of the world in such a short time. More changes occur in science and technology in a single month than formerly occurred in a century. The breathtaking pace of technological advancements ensures that few of us can keep up with the rapid change and, perhaps more important, brings to light the inability of society to keep pace with the ethics or morality behind many of these changes. These are all signs and indications that the end is rapidly approaching.

"But thou, O Daniel, shut up the words, and seal the book, even to the time of the end: many shall run to and fro, and knowledge shall be increased" (Daniel 12:4). This prophecy by Daniel from some 2,500 years ago refers to the fact that Daniel's prophecies would not be understood until the time of the end. Even in Daniel's own time, nobody had any idea of what their meaning was, and it will not be until the end that people get an actual understanding of their meaning, especially when many of these prophecies have already been fulfilled, or are in the process of being fulfilled, today.

Up until about one hundred years ago people usually never traveled more than about fifty kilometers from the place where they were born, and the fastest they could travel was the speed of a galloping horse. Only in the last fifty to one hundred years or so has travel increased to global proportions, where people, according to Daniel, are running "to and fro." The International Air Transport Association (IATA) reports that the global airline industry consists of over two thousand airlines operating more than twenty-three thousand aircraft, providing service to over thirty-seven hundred airports. Even as far back as 2006, the world's airlines flew almost twenty-eight million scheduled flights and carried well over two billion passengers from one corner of the globe to the other.

The Rapture, however, should not be confused with where the Bible speaks of the Second Coming of Christ. These are two completely different events. During the Rapture, Christ does not actually descend down to Earth; it is very clear that the saints will meet the Lord in the air, in the clouds, and this will clearly occur prior to the Tribulation. The Lord's Second Coming will occur at the end of the Tribulation Period, when Jesus physically returns to the earth at the end of the battle of Armageddon to defeat the armies of the Antichrist. The Second Coming of Christ is the most important event described in the entire Bible. The Second Coming of Christ is referred to 1,845 times in the Old and New Testaments, eight times more often than the Lord's First Coming, and

Jesus refers to His own return twenty-one times. (Matthew 16:27, Acts 1:9–11, Hebrews 9:24–28, and Revelation 2:12.)

According to a 1997 issue of *US News and World Report*, 66 percent of Americans, including a third of those who admit they never attend church, say they believe Jesus will return to Earth someday.

The Antichrist

Biblical prophecies clearly predict the coming of the Antichrist in the end times. As mankind rushes toward the end of the age, the coming of this powerful world leader is inevitable. Many believe that the end times have already started and that this individual is alive and well here on Earth today. There are more than a hundred verses and biblical passages that give a fairly accurate description about the career, nationality, character, and conquests of this demonic leader. The Antichrist, or the Beast, will indeed be the most incredible leader the world has ever known. He will make Adolf Hitler look like a heavenly angel. This evil, Satan-inspired leader will rise to dominate the world in the last days for the ultimate purpose of glorifying Satan. Daniel 7:20 tells us that he is an intellectual and theatrical genius, and Daniel 8:24–25 tells us that he is a military and a commercial genius as well. Daniel 11:21 tells us that he is a political genius of incredible magnitude; he will come to power and "seize it through intrigue." 2 Thessalonians 2:4 states that he is also a religious genius. Here seems to arrive a super intelligent political individual the world is in desperate need of in troubled times.

The New Testament does not give a clear answer as to whether the Antichrist will be a Jew or a Gentile; however, most scholars believe he will be a Gentile because his rule is part of the "Times of the Gentiles" and their authority over Israel (Luke 21:24). Daniel 7:8–24 clearly indicates that he will become the leader of the "Revived Roman Empire," or the newly formed European Union of Gentile nations, and his peace treaty with Israel will promise Gentile protection for Israel. (Daniel 9:27). Israel and the Middle East will still remain the political hot spot of the world, and this new self-appointed leader will successfully negotiate a seven-year peace treaty with the nation of Israel. However, after three and one-half years, the Antichrist will break his treaty, and then his true character will be revealed. He will control the economy of the world by issuing the mark of the Beast.

However, the Antichrist's identity will not be revealed until after the Rapture, which will cast the entire world into a state of chaos, instability, disorder, anarchy, turmoil, and collapse of financial systems, which will make our wonderful world seem to be self-destructing. Everybody will be looking for answers and solutions, and everybody will be looking for a savior—either God or the devil. The Antichrist eventually will exalt himself above God to be worshiped in the Lord's temple in Jerusalem, proclaiming himself to be God. Verses 9 and 10 say that the Antichrist will perform counterfeit miracles, signs, and wonders to gain a following and deceive many.

The time during which the world is in turmoil is the exact right time for the Antichrist to weasel his way to power, making his move to control the failing global economy and the financial and political systems of the world. He will captivate and overwhelm the population of the entire world with his superhuman intellect and his economic and military strategies. He will be viewed and treated as a hero and the true savior of our troubled world, providing all the proper answers and solutions for every problem and every question, by the people that are left behind.

During the past decades, many world leaders have been identified as possible Antichrists for various reasons, including John Kennedy, Lyndon Johnson, Henry Kissinger, Adolf Hitler, and yes, even the sweet Ronald Reagan; however, such foolish speculation should be avoided and not taken seriously. The Antichrist will not be revealed until after the Rapture. There has probably been more discussion and speculation about the Antichrist than any other Bible prophecy.

People will be eager to get the mark of the Beast because without it they will not be able to buy or sell anything. Accepting the mark of the Beast will be the only logical way to function in society; without it, no one will be able to work and get paid, because cash will no longer exist. The world is going to experience a completely electronic economy. Every financial transaction will be electronically processed. However, this technology is nothing new; the system has already been active for some forty years or more. Almost everybody nowadays possesses a credit card or a debit card; this little piece of plastic can be taken anywhere in the world and used to purchase just about anything anyone could desire, from rubber bands to automobiles. For example, if you were to make a credit card purchase in Timbuktu, Mali, let's say for 200,000 CFA (approximate value US$450), using a US dollar account, a supercomputer will instantly read the information from your card, calculate the exchange rate for the local purchase, check for sufficient funds in the account, and, if everything checks out okay, approve the purchase, and then you can bridle your camel and take it for a ride.

When the Antichrist comes to power, this is exactly what will happen. First he will require each and every one to receive the mark. There has been, and still is much speculation as to what this mark actually will consist of. It is possible that a VeriChip could be implanted in the forehead or right hand as a part of the mark of the Beast. The VeriChip is a small radio-frequency identification device. It has an identification number, is the size of a grain of rice, and is currently used in the United States and some other countries for medical information, but it could also be used for financial purposes. The truth of the matter is that no one knows precisely what system will be implemented by the Antichrist; it may very well be the VeriChip or a microchip, RFID, a barcode, just an ordinary piece of plastic, or a combination of some of these. Whatever system the Beast chooses, one thing is for certain—the technology has already been invented and is already in place.

"And he causeth all, the small and the great, and the rich and the poor, and the free and the bond, that there be given them a mark on their right hand, or upon their forehead; and that no man should be able to buy or to sell, save he that hath the mark, *even* the name of the beast or the number of his name. Here is wisdom. He that hath understanding, let him count the number of the beast; for it is the number of a man: and his number is Six hundred and sixty and six" (Revelation 13:16–18). Humanly speaking, getting the mark of the

Beast will be the logical, sane, safe, smart thing to do. But spiritually speaking, it is not wise, because accepting the mark of the Beast is like selling your soul to the devil, so to speak.

> And another angel, a third, followed them, saying with a great voice, If any man worshippeth the beast and his image, and receiveth a mark on his forehead, or upon his hand, he also shall drink of the wine of the wrath of God, which is prepared unmixed in the cup of his anger; and he shall be tormented with fire and brimstone in the presence of the holy angels, and in the presence of the Lamb: and the smoke of their torment goeth up forever and ever; and they have no rest day and night, they that worship the beast and his image, and whoso receiveth the mark of his name. (Revelation 14:9–11)

The Antichrist will be in complete control of everybody's finances. He will know exactly how much every person on Earth spends on what, where, and when. For example, he may even prevent people who have a drinking problem, or have one or more charges related to impairment on their record, from purchasing alcoholic beverages altogether by programming the supercomputer to disallow the purchase of any alcohol by that person either in a bar or in a liquor store. He may even program the supercomputer to control everybody's taxation by classifying all purchases as deductible or nondeductible and then automatically calculating taxes every month or every year, meaning that no one could even cheat on their taxes anymore. The only reason cash still exists in today's society is likely to allow criminals and drug dealers to continue to conduct their evil business ventures and allow corruption to flourish.

A global society is inevitable; this world is rapidly rushing to a world government. The concept of a world government originated with the formation of the US Federal Reserve System in 1913. The establishments of the League of Nations in 1919, the International Monetary Fund in 1944, the United Nations in 1945, the World Bank and the World Health Organization in 1948, the European Union and the euro currency in 1993, the World Trade Organization in 1998, and the African Union in 2002 are major milestones in harmony with Bible prophecy regarding a future world government. For the Antichrist, in order to control all world events, all he needs to do is gain control over these institutions. With each passing day, more and more nations are ceding their sovereignty to larger institutions, such as the European Union, the United Nations, the International Monetary Fund, and the World Court. Given the current global financial crisis and the potential catastrophe of a global war, more and more diplomatic leaders and world politicians are calling for global government. For example, in March 2009, as a result of the Asian financial crisis of 2007–2010, the People's Republic of China and the Russian Federation pressed for urgent consideration of a new international reserve currency, and the United Nations Conference on Trade and Development proposed greatly expanding the IMF's special drawing rights. Indeed the world is on a downward spiral. Never in the history of the world has there been more government corruption, financial instability, crime, fraud, and disregard for others.

Daniel's Seventy Weeks

Of all the prophetic mysteries in scripture, few have attracted the interest of Bible scholars as much as the prophecy known as the seventy weeks of Daniel. These prophecies were debated and studied even back in Daniel's own lifetime, about 2,500 years ago. The seventy weeks prophecy provides us with valuable critical information as we approach the end of the age; it helps us to connect the dots to give us an accurate timeline for those crucial events. The implication of the study of the seventy weeks of Daniel is immense and is one of the most important end-time prophecies in all of scripture. The archangel Gabriel clearly told Daniel that the seventy-week program would begin when a decree was issued to "restore and rebuild Jerusalem" (Daniel 9:25). When that critical event actually took place, the countdown began.

> And while I was speaking, and praying, and confessing my sin and the sin of my people Israel, and presenting my supplication before Jehovah my God for the holy mountain of my God; yea, while I was speaking in prayer, the man Gabriel, whom I had seen in the vision at the beginning, being caused to fly swiftly, touched me about the time of the evening oblation. And he instructed me, and talked with me, and said, O Daniel, I am now come forth to give thee wisdom and understanding. At the beginning of thy supplications the commandment went forth, and I am come to tell thee; for thou art greatly beloved: therefore consider the matter, and understand the vision. Seventy weeks are decreed upon thy people and upon thy holy city, to finish transgression, and to

make an end of sins, and to make reconciliation for iniquity, and to bring in everlasting righteousness, and to seal up vision and prophecy, and to anoint the most holy. Know therefore and discern, that from the going forth of the commandment to restore and to build Jerusalem unto the anointed one, the prince, shall be seven weeks, and threescore and two weeks: it shall be built again, with street and moat, even in troublous times. Know therefore and discern, that from the going forth of the commandment to restore and to build Jerusalem unto the anointed one, the prince, shall be seven weeks, and threescore and two weeks: it shall be built again, with street and moat, even in troublous times. And he shall make a firm covenant with many for one week: and in the midst of the week he shall cause the sacrifice and the oblation to cease; and upon the wing of abominations *shall come* one that maketh desolate; and even unto the full end, and that determined, shall *wrath* be poured out upon the desolate. (Daniel 9:20–27)

This prophecy by Daniel from some 2,500 years ago refers to the fact that Daniel's prophecies would not be understood until the time of the end. Even in Daniel's own time, nobody had any idea of what their meaning was, and it would not be until the end that people would get an actual understanding of their meaning, especially when many of these prophecies have already been fulfilled, or are in the process of being fulfilled. Verse 24 says, "… and to seal up vision and prophecy," which explains that these prophesies, or understandings, of "end things" would not be clear, or understood, until such time that these events actually took place. As is the case with many biblical prophesies, we don't know if they have been fulfilled until we realize that they have been fulfilled.

The book of Daniel was written while the Jews were exiled in Babylon because of their sin and rebellion against God. Daniel was one of the young Jewish men taken captive by the Babylonians under King Nebuchadnezzar when he conquered Israel. Daniel and some of his bright young friends (Shadrach, Meshach, and Abednego) were trained and educated for leadership positions in Nebuchadnezzar's royal court. After a three-year initiation period, they entered the king's service and were considered spiritual advisors in the same degree as the court astrologers and soothsayers, and they were judged to be "ten times better than all the magicians and enchanters in the kingdom." Daniel had a unique talent for interpreting dreams and visions. During his reign, King Nebuchadnezzar had a disturbing dream and asked his wise men to interpret it, but he refused to accept their explanations. In an uncontrollable rage of anger, he sentenced all of them, including Daniel and his friends, to death. Daniel mediated and asked for a temporary delay of execution so that he could appeal to his God for an interpretation. During the night he received a descriptive vision and explained the meaning of the king's dream the following day.

Nebuchadnezzar had a dream of an enormous statue made of four metals; the statue had a head of gold, a breast and arms of silver, a belly and thighs of bonze, legs of iron, and feet of iron and clay. Then the image was completely destroyed by a rock that turned into a huge mountain, filling the whole earth. Nebuchadnezzar's dream provided an excellent insight to world history and succeeding empires.

Daniel received a vision that interpreted the king's dream as a series of successive kingdoms. The first kingdom was Nebuchadnezzar's own kingdom of Babylon, represented by the head of gold, which reached its peak during Daniel's time. The second kingdom was Medo-Persia, which conquered Babylon in the fifth century BC. The story of the overthrow is told in the fifth chapter of Daniel, where King Nebuchadnezzar's successor, King Belshazzar, sees mysterious fingers writing on the wall. Daniel is called to read the writing, and God reprimands King Belshazzar for his pride and commits the Babylonian kingdom into the hands of his enemies, the Medes. It is from this story that we get the common phrase "The writing is on the wall."

After the second kingdom, Daniel's vision showed " …another, a third kingdom of bronze, which shall rule over all the earth." This prophecy was fulfilled approximately three centuries after Daniel's death, when Medo-Persia was attacked by Alexander the Great of Greece, who conquered the then-known world. After Alexander died in Egypt at age thirty-two, his empire was divided between four of his top generals, from which emerged the fourth world empire, the Roman Empire, represented by the legs of iron, sometime between 250 and 230 BC. At its peak, the Roman Empire covered almost the entire known Western world at that time, including most of Europe, the Middle East, Egypt, and North Africa. About AD 300, during the reign of Constantine, the empire was divided in two, forming the Eastern and Western Empires, symbolized by the two legs of Nebuchadnezzar's statue. After about three hundred years, the Roman Empire disintegrated from within and divided into a consortium of different independent nations, from which a ten-nation confederation will emerge in the form of a "Revived Roman Empire." End-time prophecies all refer to the Roman Empire as being in power immediately prior to Armageddon, and it becomes clear why Christians are looking to the European Union as the "Revived Roman Empire." "And whereas thou sawest the feet and toes, part of potters' clay, and part of iron, it shall be a divided kingdom; but there shall be in it of the strength of the iron, forasmuch as thou sawest the iron mixed with miry clay. And as the toes of the feet were part of iron, and part of clay, so the kingdom shall be partly strong and partly broken" (Daniel 2:41–42).

Then the image was completely destroyed by a rock that turned into a huge mountain, filling the whole earth. This is where God tells Daniel that this rock represents a fifth and final kingdom, a kingdom that belongs to God alone. This kingdom will arise at the finale of world history and bring the earth under God's supreme authority.

"Forasmuch as thou sawest that a stone was cut out of the mountain without hands, and that it brake in pieces the iron, the brass, the clay, the silver, and the gold; the great God hath made known to the king what shall come to pass hereafter: and the dream is certain, and the interpretation thereof sure" (Daniel 2:45). Finally all of these dominions are crushed by a kingdom that will "endure forever," meaning the kingdom of God! "I saw in the night-visions, and, behold, there came with the clouds of heaven one like unto a son of man, and he came even to the ancient of days, and they brought him near before him. And there was given him dominion, and glory, and a kingdom that all the peoples, nations, and languages should serve him: his dominion is an everlasting dominion, which shall not pass away, and his kingdom that which shall not be destroyed" (Daniel 7:7–8).

Daniel's wisdom and impeccable character gained him favor by Nebuchadnezzar, eventually elevating him to the position of second-in-command—a prime minister, so to speak. In Daniel 9:24–27, the angel Gabriel encouraged the Jewish people to end their sin and rebellion, as they were given a "second chance" to return home and rebuild Jerusalem and their temple and ultimately receive their long-promised Messiah, the Lord Jesus Christ.

The prophecy of the seventy weeks of Daniel is not really all that complicated. In the original Hebrew language, "seventy weeks" is a translation from the expression "seventy sevens." The "sevens" could refer to just about anything: days, weeks, months, or years; but after intensive study, scholars concluded that each and every one of Daniel's sevens represents a seven-year period.

The seventy-week timetable is divided into three parts. The first section is a period of 7 sevens and the second is a period of 62 sevens, totaling 69 sevens, and there is 1 seven left, each seven representing a seven-year period. This prophecy is expressed is in terms of biblical years, with each year consisting of precisely 360 days—12 months of 30 days each. The compass rose and the division of a circle are also based on this very same principal. The 360 degrees in a circle are made up of 12 segments of 30 degrees each; the biblical year is made up of 12 months of 30 days each. This is the biblical year that was issued in the quest for mathematical perfection of timekeeping.

The first section of the seventy-week prophecy represents a period of 7 weeks (or "sevens"), equaling 49 years. This period began with the edict of Artaxerxes given to Nehemiah on the first day of the Hebrew month Nisan in 445 BC. This was the proclamation that allowed the Jews to return to Jerusalem and start rebuilding the city walls and gates. It also made provisions for the reestablishment of a political and commercial system to allow Jerusalem to function as a self-sufficient city-state. The second section of the seventy-week prophecy, "threescore and two weeks," translates as 3 × 20 plus 2, or 62, sevens, equaling 434 biblical years, followed immediately after the first 7 sevens. Gabriel told Daniel that the Messiah would be "cut off and have nothing" after the 69th week, meaning Christ would be crucified at the end of the 69 weeks.

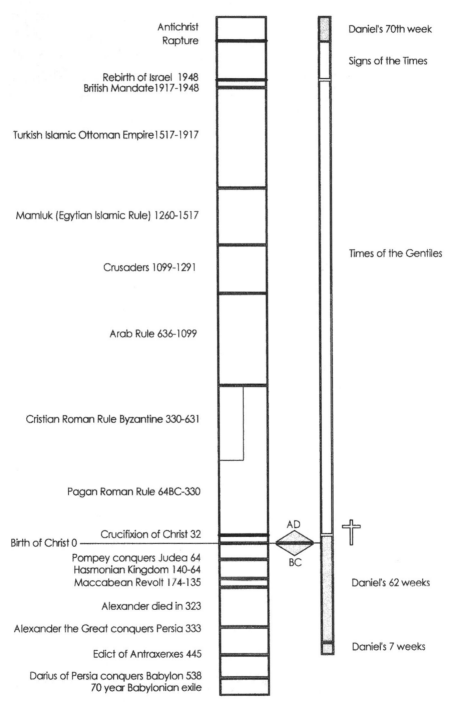

Antichrist
Rapture

Daniel's 70th week

Signs of the Times

Rebirth of Israel 1948
British Mandate1917-1948

Turkish Islamic Ottoman Empire1517-1917

Mamluk (Egytian Islamic Rule) 1260-1517

Crusaders 1099-1291

Times of the Gentiles

Arab Rule 636-1099

Cristian Roman Rule Byzantine 330-631

Pagan Roman Rule 64BC-330

Crucifixion of Christ 32

AD

Birth of Christ 0

Pompey conquers Judea 64
Hasmonian Kingdom 140-64
Maccabean Revolt 174-135

BC

Daniel's 62 weeks

Alexander died in 323

Alexander the Great conquers Persia 333

Edict of Antraxerxes 445

Daniel's 7 weeks

Darius of Persia conquers Babylon 538
70 year Babylonian exile

Time line from Babylonian captivity to the Rapture

The first two sections combined, 7 + 62 (equaling 69 sevens), totals 483 biblical years (69 × 7) or 173,880 days (483 × 360). This is the most crucial time period. We have calculated the duration of the period, but we still do not know where it will fit between two significant dates in biblical history. This was formally determined for the first time by Scotland Yard chief inspector Sir Robert Anderson (1841–1918) back in the late nineteenth century. Anderson's book *The Coming Prince*, first published in Great Britain in 1894, explained that the time period between the edict of Artaxerxes and "shall the anointed one be cut off" (the crucifixion of Christ) consisted of exactly 173,880 days. Earlier decrees under Cyrus and Ezra did not meet the criteria to initiate the prophecy of the seventy weeks.

Daniel's 69 weeks ended on Palm Sunday, the 10th day of the Hebrew month of Nisan (April 9 of AD 32), the day which saw the first coming of Christ as "Messiah the Prince" enter into His holy city through the Eastern Gate, riding on a donkey, fulfilling the prophesies of Psalm 22 and Isaiah 53. So, about two thousand years ago, significant prophetic history was made on Palm Sunday with the conclusion of Daniel's 69 weeks. "Messiah the prince" came riding into town on a donkey, about to be "cut off" (crucified). However, there is still one week remaining that still needs to be fulfilled. The reason the prophecy of Daniel's seventy weeks is so important is that it also contains crucial information regarding the end-time drama. The angel Gabriel revealed information to Daniel regarding not only the First Coming, but also the Second Coming of Christ. This final seven-year period will begin when the Antichrist signs a seven-year treaty with Israel. In fact, the revealing of the Antichrist's identity, the final merciless persecution of the Jews and Christians during the Great Tribulation, and the return of the Messiah all occur during this last seven-year period. The entire book of Revelation is devoted to the details of this terrifying time. This seven-year period is known as Daniel's seventieth week, the time of Jacob's trouble, or the seven-year Tribulation Period, which has not yet occurred.

Israel Reborn

The return of the Jews to Israel is also one of the most significant end-time Bible prophecies. Most other end-time prophecies concerning Jews pertain to the Jewish people living in Israel and not those scattered among the Gentiles. Israel had not existed as a nation since AD 70, when the Romans destroyed Jerusalem and dispersed the Jews among the nations of the world, as was predicted by the prophets. Miraculously, through an act of God, after nearly two thousand years, the nation of Israel was reborn on May 15, 1948. Never before in the history of mankind had any nation been so often defeated, persecuted and destroyed, its people scatted to the "nations of the earth," and finally been returned to the very place where it originated some four thousand years earlier with the covenant of Abraham. Whenever Israel has existed as an independent nation, conflicts have been a reality. Whether it was the Egyptians, Amalekites, Midianites, Moabites, Ammonites, Amorites, Philistines, Assyrians, Babylonians, Persians, or Romans, the nation of Israel has always been attacked and persecuted by its surrounding neighbors.

Theodor Herzl (1860–1904) initiated the prospect of the Jewish people returning to their promised land in the late nineteenth century by organizing the World Zionist Congress. The first large wave of modern immigration, known as the First Aliyah, began in 1881 as Jews fled persecution in Eastern Europe. Although the Zionist movement already existed in theory, Herzl, an Austro-Hungarian journalist, is credited with founding political Zionism, a movement that sought to establish a Jewish state in the land of Israel by elevating the Jewish question to the international level. In 1896, Herzl published *Der Judenstaat* (meaning "The Jewish State"), presenting his vision of a future Jewish state; the following year, he presided over the first World Zionist Congress. Jews slowly started to return to Palestine, causing dismay for the Arabs already living in the area. After many local confrontations, two world wars, and a considerable amount of international political manipulation, on May 14, 1948, David Ben Gurion read the Declaration of Independence and Israel was reborn as an independent state that was immediately recognized by the United States, Russia, and many other nations. The neighboring Arab nations attacked the newly created nation of Israel the very same day, pitting some forty-five million Arabs against about one hundred thousand Jews. Against incredible odds, the Jews prevailed, not only winning the war but also gaining considerable territory.

In 1956, during the Suez Canal Crisis, Israel joined a secret alliance with Great Britain and France with the intention of regaining control of the Suez Canal, which the Egyptians had nationalized and blocked for Israeli shipping. During the conflict, Israel captured the Sinai Peninsula but was pressured by the United States and the Soviet Union to withdraw in return for guarantees of Israeli shipping rights in the Red Sea and the Suez Canal.

The Six-Day War, or June War, also known as the 1967 Arab-Israeli War or the Third Arab-Israeli War, was fought between June 5 and June 10, 1967, by Israel and the neighboring states of Egypt, Jordan, and Syria. At the war's end, Israel again prevailed and seized the Gaza Strip and the Sinai Peninsula from Egypt, the West Bank and East Jerusalem from Jordan, and the Golan Heights from Syria.

The Yom Kippur War, also known as the October War, began on October 6, 1973, the Jewish Day of Atonement, the holiest day in the Jewish calendar and a day when adult Jews are required to fast. The Syrian and Egyptian armies launched a well-executed surprise attack against the unsuspecting Israeli Defense Forces. For the first few days there was a great deal of doubt about Israel's capacity to resist the invading armies; however, the Syrians were forced back, and although the Egyptians captured a strip of territory in Sinai, Israeli forces had in turn crossed the Suez Canal and advanced to only one hundred kilometers from Cairo, the capital of Egypt.

Although the war's results were generally favorable to Israel, it cost more than two thousand Jewish lives and resulted in a heavy arms debt. The war generally made Israelis more aware of their vulnerability. Following the war, both Israelis and Egyptians showed greater willingness to negotiate for peace.

Israel is still a political hot spot in the world today, and it continues to be persecuted by its age-old enemies the Palestinians, Syria, Lebanon, Jordan, Saudi Arabia, Iran, Hamas, Islamic Jihad, and Hezbollah, among many others. But this hatred and persecution of Israel gives us only a slight indication of what we can expect to happen in the end times (Matthew 24:15–21). Many Bible prophecy scholars believe the six-day Arab-Israeli war of 1967 marked the "beginning of the end."

The rebirth of Israel during the "last generation" is indeed one of the greatest prophetic miracles of all times, but how does the last generation fit the equation? Let us examine what the Bible has to say about a generation: "And Jehovah's anger was kindled against Israel, and he made them wander to and fro in the wilderness forty years, until all the generation, that had done evil in the sight of Jehovah, was consumed" (Numbers 32:13). This clearly defines a generation as a forty-year period; however, Psalm 90:10 states, "The days of our years are threescore years and ten, Or even by reason of strength fourscore years; Yet is their pride but labor and sorrow; For it is soon gone, and we fly away," defining a generation as a seventy- to eighty-year period. "And Jehovah said, My spirit shall not strive with man forever, for that he also is flesh: yet shall his days be a hundred and twenty years" (Genesis 6:3).

Some biblical scholars feel that the generation that Jesus referred to was present when Israel was reborn as an independent nation. Therefore, the end of this generation in this perspective could be considered 1988, 2018, or 2068. However, a generation in today's world is considered to be somewhere between seventy and eighty years. Is it therefore not reasonable to accept that we can expect the Rapture to occur somewhere before 2021 (1948 + 80 - 7)?

There are many prophecies predicting the restoration of Israel in one day (May 15, 1948). For example:

> For I will take you from among the nations, and gather you out of all the countries, and will bring you into your own land ... Thus saith the Lord Jehovah: In the day that I cleanse you from all your iniquities, I will cause the cities to be inhabited, and the waste places shall be builded. And the land that was desolate shall be tilled, whereas it was a desolation in the sight of all that passed by. And they shall say, this land that was desolate is become like the Garden of Eden; and the waste and desolate and ruined cities are fortified and inhabited. (Ezekiel 36:24, 33–35)
>
> After many days thou shalt be visited: in the latter years thou shalt come into the land that is brought back from the sword, that is gathered out of many peoples, upon the mountains of Israel, which have been a continual waste; but it is brought forth out of the peoples, and they shall dwell securely, all of them. (Ezekiel 38:8)

Moses predicted some 3,500 years ago that the Jews would be dispersed among the nations of the world and would be persecuted, precisely as did happen during the last two thousand years, and that one day the Jews would return to God and to the land of Israel, which He promised to Abraham, Isaac, and Jacob.

> And Jehovah will scatter you among the peoples, and ye shall be left few in number among the nations, whither Jehovah shall lead you away ... When thou art in tribulation, and all these things are come upon thee, in the latter days thou shalt return to Jehovah thy God, and hearken unto his voice: for Jehovah thy God is a merciful God; he will not fail thee, neither destroy thee, nor forget the covenant of thy fathers which he sware unto them. (Deuteronomy 4:27, 30–31)

Jehovah did not set his love upon you, nor choose you, because ye were more in number than any people; for ye were the fewest of all peoples: but because Jehovah loveth you, and because he would keep the oath which he sware unto your fathers, hath Jehovah brought you out with a mighty hand, and redeemed you out of the house of bondage, from the hand of Pharaoh king of Egypt. (Deuteronomy 7:7–8)

For, lo, the days come, saith Jehovah, that I will turn again the captivity of my people Israel and Judah, saith Jehovah; and I will cause them to return to the land that I gave to their fathers, and they shall possess it. (Jeremiah 30:3)

And it shall come to pass in that day, that the Lord will set his hand again the second time to recover the remnant of his people, that shall remain, from Assyria, and from Egypt, and from Pathros, and from Cush, and from Elam, and from Shinar, and from Hamath, and from the islands of the sea. (Isaiah 11:11)

For, behold, in those days, and in that time, when I shall bring back the captivity of Judah and Jerusalem, I will gather all nations, and will bring them down into the valley of Jehoshaphat; and I will execute judgment upon them there for my people and for my heritage Israel, whom they have scattered among the nations: and they have parted my land. (Joel 3:1–2)

In that day will I raise up the tabernacle of David that is fallen, and close up the breaches thereof; and I will raise up its ruins, and I will build it as in the days of old. (Amos 9:11)

Earthquakes and the Bible

> For nation shall rise against nation, and kingdom against kingdom; and there shall be famines and earthquakes in divers places. (Matthew 24:7)

> For nation shall rise against nation, and kingdom against kingdom; there shall be earthquakes in divers places; there shall be famines: these things are the beginning of travail. (Mark 13:8)

> ...and there shall be great earthquakes, and in divers places famines and pestilences; and there shall be terrors and great signs from heaven. (Luke 21:11)

According to a number of Christian writers and teachers on Bible prophecy, including Hal Lindsey, Chuck Missler, Jack Van Impe, J. R. Church, Grant Jeffrey, Gary Stearman, John Hagee, and Peter and Paul Lalonde, Jesus predicted in the Olivet Discourse that a pronounced increase in the frequency and intensity of earthquakes would occur just prior to His return. For example, Hal Lindsey, the world's best-known teacher of biblical prophecy and author of seventeen books on the subject, wrote in his 1997 book *Apocalypse Code*: "Earthquakes continue to increase in frequency and intensity, just as the Bible predicts for the last days before the return of Christ. History shows that the number of killer quakes remained fairly constant until the 1950s - averaging between two to four per decade. In the 1950s, there were nine. In the

1960s, there were 13. In the 1970s, there were 51. In the 1980s, there were 86. From 1990 through 1996, there have been more than 150"

These numbers have been extensively publicized by popular prophecy teachers such as Chuck Missler and Jack Van Impe. In his book *Planet Earth 2000 A.D.*, Lindsey cites the United States Geological Survey (USGS) in Boulder, Colorado, as the source for this information but does not specify any specific report or paper or statistics.

Grant R. Jeffrey, another Bible prophecy teacher and author of several best-selling books, could very well be the source of Lindsey's statistics. Two years before Lindsey's statement was published, Jeffrey wrote, "However, since A.D. 1900, the growth in major earthquakes has been relentless. From 1900 to 1949 it averaged three major quakes per decade. From 1949 the increase became awesome with 9 killer quakes in the 1950's; 13 in the 60's; 56 in the 1970's and an amazing 74 major quakes in the 1980's. Finally, in the 1990's, at [sic] the present rate, we will experience 125 major killer quakes in this decade" (Jeffrey 1994). Jeffrey also cites the US Geological Survey Earthquake Report, Boulder, Colorado, as the source, but also without any further specifications.

John Hagee is the founder and pastor of the fifteen-thousand-member Cornerstone Church in San Antonio, and in 1995 he wrote the book *Beginning of the End*, which became a *New York Times* bestseller. In this book Hagee refers to a report from the National Earthquake Information Center of the US Geological Survey: "...the number of large earthquakes (magnitude 6.0 or greater), have stayed relatively constant ... the last decade has produced substantially fewer large earthquakes than shown in the long-term averages" (Hagee 1996).

Remarkably, Hagee goes on to directly contradict the government report with the following statement: "It is true that the Bible predicts that earthquakes will increase in the last days, and the number of earthquakes measured has in-creased 1.58 times between 1983 and 1992." The documentation produced to support the supposed increase is faulty. Adequate reason is not given as to why the conclusion of the government report (i.e., a decreasing frequency of earthquakes) should be rejected and ignored.

Steven A. Austin is chairman of the Geology Department at the Institute for Creation Research in Santee, California, and Mark L. Strauss is associate professor of New Testament at Bethel Seminary in San Diego, California. These professional individuals certainly qualify as experts in biblical references to earthquakes. They tend to disagree with the statements of the above writers and published an extensive commentary on their website, on which this chapter is based.

The truth of the matter is that, contrary to what prophecy teachers tell us, there is no obvious trend indicating an abnormal increase in the frequency of large earthquakes during the last half of the twentieth century. Neither is there a noteworthy decline of earthquakes in the first half of the century. The decades of the 1970s, 80s, and 90s rather experienced a decline in larger earthquakes compared to earlier decades of the century. The 70s, 80s, and 90s are precisely those decades that many prophecy teachers target, incorrectly showing a dramatic increase of larger earthquakes. Regional data from earthquake hotspots like California and Japan also do not indicate an increase in earthquake frequency in the latter part of the last century.

A partial explanation may lie in the fact in 1931, there were about 350 seismograph stations operating around the globe; today there are more than 8,000, and the data from these stations is instantly transmitted by electronic mail, Internet, and satellite. This increase in the number of stations and improved data management has allowed us to locate earthquakes more quickly and to locate many smaller quakes, which remained undetected in earlier years. Because of the improvements in communications and also an increased interest in environment and natural disasters, the public is now much more aware of earthquakes than in years past.

The fact of the matter is that there has been a decline in earthquake frequency through the 1960s (20.4 events per year), 1970s (20.4 events per year), and 1980s (11.2 events per year). Those are the decades Hal Lindsey, Grant Jeffrey, and John Hagee proclaim that the frequency of big earthquakes was increasing. However, from 1990 through 1997, there has been an average of 17.3 big earthquakes per year, which is still well below the average of 20.0 quakes per year for the entire century.

If earthquakes are not on the increase, then how should we interpret Matthew 24:7? As noted above, Hal Lindsey implies that earthquakes will continue to increase "just as the Bible predicts for the last days." Closer examination of the New Testament evidence reveals that Lindsey's statement is wrong on both counts. Not only are earthquakes not increasing, but also the biblical text never indicated that they would. The popular conception that an increase of earthquakes in frequency and severity is a "key sign" of the nearness of the Second Coming of our Lord is simply a result of misinterpretation of the biblical text.

There are good reasons not to believe that Jesus's words indicate an increase in frequency or severity of these general signs, but rather that they indicate only that the signs will continue until the end of the age. He rather downplayed their significance and encouraged his followers not to be alarmed or disturbed by them. He certainly did not indicate that their frequency should be counted in an attempt to calculate or predict a precise time for the end.

Earthquakes and other cataclysmic events are significant in scripture, demonstrating the awesome power of Almighty God. At Mount Sinai the Lord's presence was indicated by smoke and the shaking of the mountain (Exodus 19:18; Psalm 68:8; Job 9:6). When the New Testament church prayed, "the place where they had gathered together was shaken," and the Spirit's presence was manifested (Acts 4:31). Paul and Silas were freed from prison when God's power and presence was manifested in an earthquake (Acts 16:26).

The most unusual earthquakes were associated with the crucifixion and resurrection of Christ. When Christ died on the cross, an earthquake shook the temple and rent the curtain of the temple from top to bottom. (Matthew 27:51). No human being could have rolled away the stone that sealed Christ's tomb; it was the angel in the midst of the earthquake (Matthew 28:2). More specifically, many seismic events are manifestations of God's anger and righteous judgment (1 Samuel 14:15; Psalm 18:7–8; Isaiah 5:25; 13:13; 29:6; Joel 3:16; Amos 1:1, 2; 8:7, 8; Micah 1:3–7; Nahum 1:5, 6; Haggai 2:6, 21).

In the book of Revelation, earthquakes are symbols of God's final judgment upon the earth. Five earthquakes are described. These occur at the opening of the sixth and seventh seals (Revelation 6:12, 8:5), just before and after the seventh trumpet (Revelation 11:13, 19) and during the seventh bowl (Revelation16:18). This last earthquake is identified as the worst ever, splitting Jerusalem into three parts and destroying the cities of the nations. Although they demonstrate the awesome power and presence of our Lord, these passages do not indicate an increase in earthquakes in the present age. For those who follow a futuristic and dispensational interpretation of Revelation, these earthquakes are predicted to happen during the Great Tribulation, not prior to it.

Corruption

Evil, violence, and corruption have been a reality ever mankind has existed. Adam and Eve were the first humans created by God, who placed them in the garden of Eden. Eve was deceived by the serpent, and she ate of the forbidden fruit. Then Adam, who was with her, chose to eat of the fruit also, and they committed the first sin, disobeying the commandments of God. This event represents the first act of corruption in the Bible. Eve and Adam practiced dishonesty and tried to do things their way. When they sinned, spiritual and physical death came into the world and fellowship with God was broken.

Cain and Abel were, according to the book of Genesis, the first two sons born to Adam and Eve. Cain is described as a crop farmer and his younger brother Abel as a shepherd. Cain was the first human born, and Abel was the first human to be murdered by the hand of his brother. Modern interpreters typically assume that the motive for the murder was jealousy and anger due to God's acceptance of Abel's offering and His rejection of Cain's.

Only ten generations after Adam and Eve, humanity had become evil, violent, and corrupt to the point where God decided to wipe out the whole mess—the Earth and all humanity—except for one righteous man and his family, Noah. Because of the people's wickedness and disobedience, God gave Noah a warning: "As for me, here I am bringing the deluge of waters upon the earth to bring to ruin all flesh" (Genesis 6:7). Noah, a "preacher of righteousness," tried to make the people aware of this coming catastrophic event, but they completely ignored his warnings and kept perusing their evil ways. Then Noah was instructed by God to "make an ark" and fill it with two of every sort of living things, as well as his family. The book of Genesis indicates that God intended to return the earth to its pre-Creation state of chaos by flooding the earth for 370 days—150 days of flooding and 220 days to dry up the floodwaters. After the global destruction of the flood, God made a covenant with Noah and his family, promising them "that never again shall all flesh be cut off by the waters of the flood, and never again shall there be a flood to destroy the earth."

Sodom and Gomorrah deliver another example of God's punishment of an evil and corrupt society. The story of Sodom and Gomorrah is not a fairy tale. According to archeologist Ron Wyatt, things happened exactly as the biblical accounts present them, except it was evil, corruption, and sexual perversion rather than wickedness and violence that caused the destruction. This time it was fire and brimstone that caused annihilation rather than a flood, which God had promised would never happen again. There is a lesson to be learned here. Today's society behaves much like the people in the days of Noah, with an abundance of evil, violence, and corruption. Today is also the time when Matthew 24:37–39 requires a closer examination: "But as the days of Noah were, so shall also the coming of the Son of man be. For as in the days that were before the flood they were eating and drinking, marrying and giving in marriage, until the day that Noe entered into the ark, and knew not until the flood came, and took them all away; so shall the coming of the Son of man be" (KJV). Our society today behaves much like previous evil, violent, and corrupt societies. Homosexuality and sexual perversion are accepted or condoned in most Western nations, and the news media focuses mainly on murders and political corruption.

In the second book of Samuel we read of a particularly wicked sin of King David, the most celebrated historical king of the Israelites. David commits adultery with Bathsheba, the wife of his army commander Uriah the Hittite. Bathsheba becomes pregnant, and David summons Uriah to sleep with his wife in order to conceal the identity of the real father of the baby. Uriah refuses to comply with David's wishes because he refuses to abandon his colleagues at the siege of Rabbah. David then sends Uriah to report to Joab, his commander, with a message instructing Joab to abandon Uriah on the battlefield, " ...that he may be struck down and die." After Uriah becomes a casualty on the battlefield, David marries Bathsheba and she gives birth to his child.

Never in the history of the world has there been more government corruption, financial instability, crime, fraud, and disregard for others. Much of history connects corruption with the decline of civilizations. Political corruption happening in the world is considered one of the most serious problems in society today and is also a key indicator that the end is approaching. In its most recent report on global bribery, the Berlin-based organization Transparency International found that the demand for bribes by government officials occurs everywhere, but the highest rates of corruption are in Africa, the Newly Independent States (Russia, Ukraine, and Moldova, for example), Asia-Pacific, and Latin America.

The 2010 Corruption Perceptions Index is based on thirteen independent surveys. However, not all surveys included all countries. A score relates to perceptions of the degree of corruption as seen by businesspeople and country analysts, and rates between 10 (least corruption) and 0 (most corruption). The results indicated that Somalia was the most corrupt nation in the world, while Denmark, New Zealand, and Singapore tied for first place with scores of 9.3, meaning they were seen as having the least amount of corruption. The Netherlands came in at the eleventh spot with a rating of 8.6, Canada at number fourteen with an 8.4, The United States in seventeenth place with 7.6, and Israel in the twenty-eighth spot with a score of 6.3. Of the 160 countries surveyed, the average rating was 4.1.

Humankind has tried just about every conceivable form of government—monarchy, dictatorship, democracy, and all sorts of socialistic programs. Except for monarchies, they all seem to have disintegrated or failed within a period of about two hundred years. Winston Churchill once said, "It has been said that democracy is the worst form of government, except all the others that have been tried." None of the humanly-devised governments on Earth are perfect. They are filled with corruption, inefficiency, inconsistency, and, often, outright injustice. Men may come to power in many different ways, including favoritism, bribery, revolution, deceit, or even murder. Regardless of how leaders conduct themselves, either in their personal lives or in carrying out their official duties, they must be obeyed. But God commands us to always keep the laws of men—unless they are contrary to *His* laws, directly or indirectly. Romans 13:1 clearly states, "Let every person be in subjection to the governing authorities. For there is no authority except from God, and those which exist are established by God."

The Gettysburg Address, Abraham Lincoln's most famous speech (273 words) and one of the most quoted political speeches in US history, was delivered at the dedication of the Soldiers' National Cemetery in Gettysburg, Pennsylvania, on November 19, 1863, during the American Civil War, four and a half months after the Battle of Gettysburg. In his speech Lincoln stated, "…that these dead shall not have died in vain; that this nation, under God, shall have a new birth of freedom; and that government of the people, by the people, for the people, shall not perish from the earth." After his speech, people applauded, cheered, and rejoiced.

In his inaugural address in Washington, DC, on January 20, 1961, almost a hundred years after Lincoln's famous speech, John F Kennedy (1917–1963) also spoke words that became just as famous; they also made an impact on the entire world and will also never be forgotten. Kennedy stated, "And so, my fellow Americans: ask not what your country can do for you, ask what you can do for your country." After his speech, people applauded, cheered, and rejoiced. However, there is a fundamental difference between what Lincoln said in his speech and what Kennedy had to say about one hundred years later. Lincoln spoke of a government by the people, for the people—a government to serve the nation, so to speak. Kennedy's speech focused on the nation serving the government instead of the government serving the nation. Lincoln was talking about true democracy; Kennedy talked about what could be classified as an elected dictatorship. Obviously, a government of the people, by the people, and for the people had already perished from the earth. After the Rapture, the Antichrist will, through intrigue and manipulation, take control of the nations of the world, and again everybody will applaud, cheer, and rejoice.

It is critically important to fully contemplate, recognize, and understand that all of these changes are occurring today and have been occurring at an ever-increasing rate. The social, political, and economic changes are designed to prepare the world for the Beast, or the Antichrist. The technological changes will enable the Beast to gain control over buying and selling.

Abuse of Church and Political Authority

The history of the Catholic Church and the Popes is complicated and highly controversial. Throughout history, the Papacy has truly been a fascinating subject, is rather complexed and much to the embarrassment of Catholic believers, comes with a rather checkered and turbulent past. There have been many scandalous as well as colorful events in its past. It was not very long after the establishment of the Church before its leadership became involved in a struggle for political dominance, money and power.

Saeculum obscurum, the Latin word for "dark age", is a name assigned to this period in the history, beginning with the installation of Pope Sergius III in 904 and lasting for sixty years until the death of Pope John XII in 964. Much of what is known about the saeculum obscurum comes from the recorded histories of Liutprand, bishop of Cremona and the *Liber Pontificalis* (Book of Popes). Liutprand took part in the Assembly of Bishops, which deposed Pope John XII in 964 and was a political enemy of the Roman aristocracy. He is described by the Catholic Encyclopedia as "ever a strong partisan and frequently unfair towards his adversaries." Theodora was characterized by Liutprand as a "shameless whore ... [who] exercised power on the Roman population like a man." Her daughter Marozia (890-936), also became a Roman noblewoman who was also given the titles senatrix as well as Patricia (elite Family) of Rome by one of her mother's powerful relatives Johannus, deacon of Bologna. In 914, Theodora persuaded Theophylact to install Johannus to the Papal Throne; naming him Pope John X. Marozia was a Roman noblewoman who was the mistress of Pope Sergius III when she was fifteen years old. She was also alleged to be the lover of Pope John X, whom she had murdered in 928. Her illegitimate son she had with Pope Sergius III became Pope John XI in 931. Marozia had two lovers, a son, a nephew, one grandson, two great-grandsons and a great-great-grandson that were all pope at one time or another over a period of about 140 years. Wow! Quite some genealogy!

The story of Pope Boniface VIII (1294–1303) and King Philip IV (1285–1314) of France represents one of the most dramatic clashes between the forces of the Catholic Church and state. The conflict between Boniface and Philip took place between 1295 and 1303. With Pope Boniface VIII, the Church ended a period of reform and consolidation of power that dated roughly from the reign of Gregory VII (1073–1085). The war between France and England brought increased taxation of Church revenues. Arguments over taxation turned into a heated clash between the pope and the king, challenging the very nature of imperial and papal power. The feud between the two reached its peak when Philip began to launch a strong antipapal campaign against Boniface. In retaliation, on November 18, 1302, Boniface issued one of the most important papal bulls in Roman Catholic history: Unam sanctam. It declared that both spiritual and temporal powers were under the pope's jurisdiction and that kings were subordinate to the power of the Church. The pope, once merely the bishop of Rome, was now more of an emperor who claimed both spiritual and secular dominion over all Christianity and Christian nations. However, Philip did not see things that way and politely sort of told the pope to go fly a kite. Guillaume de Nogaret, Philip's advisor and chief minister, denounced Boniface as a heretical criminal (and practitioner of sodomy) to the French clergy. Consequently, in 1303, both Philip and Nogaret were excommunicated. In retaliation, on September 7, 1303, an army of about two thousand men led by Nogaret and Sciarra Colonna of the powerful Colonna family surprised Boniface at his retreat in Anagni, about fifty kilometers southeast of Rome. The king and the Colonnas demanded the pope's resignation, to which Boniface responded that he would "sooner die." In response, Colonna hit Boniface in the face, a slap that is still remembered and celebrated in the local tradition of Anagni today. A fistfight broke out in which Boniface was beaten badly and nearly died. Then he was held captive and deprived of food and water, but he was released after three days. After a month, on October 11, 1303, Boniface died of kidney stones and humiliation. There were also rumors he had died of suicide from "gnawing through his own arm" and bashing his skull into a wall, but some believed that he died as a result of injuries received during the beating. Nicola Boccasini was elected pope in 1303 as Benedict XI. After a short term of only eight months, Benedict XI died suddenly at Perugia. It was originally suspected that his sudden death was caused by poisoning through the organization of Nogaret.

Bertrand de Got was elected and called himself Pope Clement V (1305–1314). Clement was not Italian or even a cardinal and had never set foot in Rome. After his election, he initially relocated the papal seat from the ancient city of Rome to Poitiers (a city in the west central part of France), and four years later he moved it to the fortified city of Avignon, which was not part of France at that time. He and the French popes who succeeded him remained completely under the control and influence of Philip. Early in 1306, Clement explained away the papal bulls that seemed to apply to Philip and essentially withdrew Unam sanctam, which was particularity offensive to Philip's ambitious rule. Clement conducted himself as a puppet or a subject of the French monarchy, causing a radical change in papal policy.

In 1306 Philip expelled all Jews from France, and in 1307 he annihilated the order of the Knights Templar. Philip was heavily indebted to both groups and viewed them as enemies of his nation. He chose to eliminate the debtors rather than pay his debt to them.

On Friday, October 13, 1307, Philip arrested the last grand knight of the Knights Templar, Jack de Molay, and hundreds of his knights throughout France, to relieve the huge debt owed to the Templars, who financed his war with England. The king falsely accused the Templars with Christian heresy, misconduct for sexual perversion, idol worship, and worshipping Satan, and they were tortured into false confessions. However, the Templars were answerable to the pope only; Philip ignored the jurisdiction and dealt with the problem prior to the pope's response. Jacques de Molay and Geoffroi de Charney (the last known grand master of the Knights Templar) were burned to death on a slow fire outside Notre Dame in Paris on Friday, March 13, 1314. Jacques de Molay did not go to his God with peace. As he slowly burned, he cursed King Philip IV and Pope Clement V, calling on them to meet him before God within one year so they could all be judged together. As it turned out, both men, Pope Clement V and King Philip, both relatively young, died in the same year, 1314. Pope Clement was fifty-four years of age, while Philip was only forty-six when he died in a hunting accident. The deaths of King Philip and Pope Clement V within a year led many to fear the number thirteen—Friday the thirteenth in particular. For centuries, Friday the thirteenth has had the stigma of being an unlucky day. Well, it certainly was not a very lucky day for Jacques de Molay.

Pope Pius II commented on the conspiracy corruption that preceded his own election in 1458: "The richer and more powerful members begged, promised, threatened, and some, shamelessly casting aside all decency, pleaded their own cause and claimed the papacy as their right. Their rivalry was extraordinary, their energy unbounded. They took no rest by day or sleep by night." Several years later, Rodrigo Borgia nearly bankrupted himself to become Pope Alexander VI (1492–1503). Borgia bribed his fellow cardinals with bags of bullion, money he had extorted selling pardons for all sorts of crimes, and he could barely contain his excitement when he won. "I am pope, I am pope," he exclaimed as he dressed himself in his extravagant new vestments. However, he also won the award for worst pope ever. In a move that created much scandal, Alexander created twelve new cardinals. Among the new cardinals was his son Cesare, then only eighteen years old. Alessandro Farnese (later Pope Paul III), the brother of one of the pope's mistresses, the beautiful Giulia Farnese, was also among the newly created cardinals.

Thanks to his convenient social status, Rodrigo Borgia passed through the ranks of bishop, cardinal, and vice-chancellor, gaining enormous wealth along the way. Alexander was so corrupt that his surname eventually became a byword representing the hellishly low papal standards of the time. He fathered at least seven different illegitimate children by his mistresses and did not hesitate to reward them with generous donations at the church's expense. When low on finances, he simply established new cardinals in return for payments or slammed wealthy people with completely fabricated charges, jailed or murdered them for said false charges, and then confiscated their properties and money. Not surprisingly, there is very little about Alexander VI that can be considered godly or even lawful. His goals were selfish and ambitious, and the orderly government he initially administered quickly deteriorated until the city of Rome was in a state of complete disrepair. The words spoken by Giovanni de Medici (the future Pope Leo X) after Borgia's election are telling: "Now we are in the power of a wolf, the most rapacious perhaps that this world has ever seen. And if we do not flee, he will inevitably devour us all."

It was these imperial pretensions that made the papacy a much bigger prize. Accordingly, the papal succession became even more hostile as corrupt and greedy cardinals grasped for it, resorting to intimidation, bribery, and even murder. From mistresses and illegitimate children to dance parties and harbored criminals, the Vatican has a shockingly dirty history.

The Roman Catholic Church teaches that once an ordained priest blesses the bread, or the host, during the Holy Eucharist, it is transformed into the actual flesh of Christ, although it retains the appearance, odor, and taste of bread. When the priest blesses the wine, it is transformed into the actual blood of Christ, though it retains the appearance, odor, and taste of wine.

The Holy Eucharist is the oldest practice of Christian worship, as well as the most distinctive; it represents the celebration of the life, death, and resurrection of our Lord Jesus Christ. The word "Eucharist" is derived from a Greek word meaning "thanksgiving." The origin of the Eucharist is traced to the Last Supper in the Upper Room, where Christ instructed His disciples to offer bread and wine in His remembrance on the day prior to His suffering and crucifixion and death.

Jesus came to Earth for one purpose and one purpose only. He came here for the redemption of the sins of humanity; He came here to save the soul of every sinning human being from damnation. The Last Supper was a final get-together of our Lord and His disciples to celebrate the completion of the first part of His mission—the fulfillment of His teachings and the performance of His miracles. Through His miracles He proved to the world that He was truly the promised Messiah as spoken of in the Old Testament. The only thing that was left for Him to be completed was His suffering and sacrificial death on the cross, which He knew was going to happen the very next day; the apostles, however, had no idea. Just one more time, for the last time, Jesus wanted to remind His disciples of the importance of the purpose of His coming to Earth; just one more time Jesus wanted to remind His disciples that He, "The bread of heaven," was going to sacrifice His body, His flesh and blood, for the redemption of the sins of mankind. "… and the bread that I shall give is my flesh, which I shall give for the life of the world" (John 6:51). Jesus wanted the Last Supper to be a celebration to be remembered and commemorated for future generations, to relentlessly remind each and every human being of His sacrificial death on the cross for them.

While sharing the final supper, Jesus took bread, blessed and broke it, and shared it with His apostles, and He spoke the words "Take and eat; this is My body." Then He took a cup of wine, blessed it, gave it to them, and said, "All of you drink of this; for this is My blood of the new covenant which is being shed for many unto the forgiveness of sins." Then He specifically instructed His apostles to "Do this in remembrance of Me," indicating this was a ceremony that must be continued and repeated for future generations. The Catholic Church interprets the words "This is my body" and "This is my blood" literally, word for word, and teaches that when the bread (host) and wine are consecrated in the Eucharist, they are no longer bread and wine but miraculously change into the actual physical material body, the flesh and the blood of our Lord Jesus Christ, combined with His soul and divinity. In other words, the host actually transforms from a little wafer of bread to a complete, living human being. This is the process referred to by the Church as "Transubstantiation."

For the last ten or more centuries, transubstantiation has been a controversial issue, and it is considered one of the major causes for the Reformation. After intensive studies of the scriptures, reformers, such as William Tyndale, Jan Hus, John Calvin, Martin Luther, John Wycliffe, Menno Simons, Huldrych Zwingli, and especially Desiderius Erasmus, as well as many others, independently concluded that the Catholic Church's transubstantiation consisted of false doctrine, religious malpractice, and a complete contradiction to biblical teaching. They furthermore expressed their concerns about other Church doctrine, such as the existence of purgatory, the sale and abuses of indulgences (simony), and systematic corruption of the Church's hierarchy, including the popes. In retaliation to reformers' oppositions, in its thirteenth session, ending October 11, 1551, The Council of Trent summarizes the Catholic faith by declaring: "Because Christ our Redeemer said that it was truly his body that he was offering under the species of bread, it has always been the conviction of the Church of God, and this holy Council now declares again, that by the consecration of the bread and wine there takes place a change of the whole substance of the bread into the substance of the body of Christ our Lord and of the whole substance of the wine into the substance of his blood. This change the holy Catholic Church has fittingly and properly called transubstantiation."[206] (Catechism of the Catholic Church 1376) Therefore, many Bible scholars believe that the Church's transubstantiation decree of 1551 was largely based on political initiatives rather than biblical truth.

Is such a concept truly biblical? Some scriptures, if interpreted strictly literally, do indeed seem to point toward the "real presence" of Christ in the bread and the wine, for example in Matthew 26:26: "And as they were eating, Jesus took bread, and blessed, and brake it; and he gave to the disciples, and said, Take, eat; this is my body." Luke 22:19 says, "And he took bread, and when he had given thanks, he brake it, and gave to them, saying, This is my body which is given for you: this do in remembrance of me." We then read in 1 Corinthians 11:24–25, "And when he had given thanks, he brake it, and said, this is my body, which is for you: this do in remembrance of me. In like manner also the cup, after supper, saying, this cup is the new covenant in my blood: this do, as often as ye drink *it*, in remembrance of me." The passages Catholic believers most frequently refer to and find most specific and explicit are located in the sixth chapter of the book of John 6, especially verses 53–56. "He that eateth my flesh and drinketh my blood hath eternal life: and I will raise him up at the last day. For my flesh is meat indeed, and my blood is drink indeed" (John 6 54–55). These were actual words spoken by Jesus, and strictly based on these verses, the Catholic Church insists transubstantiation is scriptural, literally word for word, and they correlate this message with the Lord's Last Supper. These verses sound compelling indeed and even appear to suggest cannibalism. Sounds like Jesus avails himself to be butchered and put on the barbeque, so to speak. It also sounds very revolting! Did Jesus actually offer His body for human consumption? Is this what He had in mind when He spoke these words? Or, perhaps, did He disclose another parable, as He did so many times in the past when some potentially confusing event had to be explained? In order to find the actual truth regarding what Jesus really intended these words to mean, the sixth chapter of John requires extensive and serious consideration. In order to get a clear picture and the true meaning of what Jesus was saying, we must decipher the context of the whole teaching, starting with John 6:32–35: "Jesus therefore said unto them, Verily, verily, I say unto you, It was not Moses that gave you the bread out of heaven; but my Father giveth you the true bread out of heaven. For the bread of God is that which cometh down out of heaven, and giveth life unto the world. They said therefore unto him, Lord, evermore give us this bread. Jesus said unto them. I am the bread of life: he that cometh to me shall not hunger, and he that believeth on me shall never thirst."

Then, starting with verse 47, Jesus explains what He was referring to.

> Verily, verily, I say unto you, He that believeth hath eternal life. I am the bread of life. Your fathers ate the manna in the wilderness, and they died. This is the bread which cometh down out of heaven, that a man may eat thereof, and not die. I am the living bread which came down out of heaven: if any man eats of this bread, he shall live forever: yea and the

bread which I will give is my flesh, for the life of the world. The Jews therefore strove one with another, saying, How can this man give us his flesh to eat? Jesus therefore said unto them, Verily, verily, I say unto you, except ye eat the flesh of the Son of Man and drink his blood; ye have not life in yourselves. He that eateth my flesh and drinketh my blood hath eternal life: and I will raise him up at the last day. For my flesh is meat indeed, and my blood is drink indeed. He that eateth my flesh and drinketh my blood abideth in me, and I in him. As the living Father sent me, and I live because of the Father; so he that eateth me, he also shall live because of me. This is the bread which came down out of heaven: not as the fathers ate, and died; he that eateth this bread shall live forever. These things said he in the synagogue, as he taught in Capernaum. Many therefore of his disciples, when they heard *this*, said, this is a hard saying; who can hear it? (John 6:47–60)

When Jesus spoke the words "This is the bread which came down from Heaven" in verse 58, the word "This," obviously means He is referring to Himself, sent by Almighty God. And in verse 51 ("… and the bread that I shall give is my flesh, which I shall give for the life of the world"), He refers to the offering of His body to be sacrificed for the redemption of the sins of humanity. When He spoke the words "He who eats this bread will live forever," also in verse 58, He obviously meant that whoever accepts Him and believes in Him as the Savior "will live forever," meaning that a person who believes in Him will be saved and have everlasting life. Verse 47 makes it clear that whoever believes in Him shall have everlasting life. He did not say that whoever eats Him shall have everlasting life!

At least three times, Jesus makes a comparison between Himself, as "the bread which came down from heaven," and the bread Moses miraculously provided some 1500 years earlier for the Israelites during their forty-year journey through the Sinai wilderness in the form of manna. God provided manna from heaven as physical food to prevent the Israelites from hunger and starvation. God sent and sacrificed His

Son, Jesus, or "the bread from heaven," as spiritual food to prevent the souls of humanity from damnation. Manna had nothing to do with redemption of sin and everlasting life. Jesus or "the bread from heaven" had nothing to do with physical food, hunger, and starvation. These verses are obviously symbolic and figurative terms to explain a spiritual relationship, not a physical one. It was clearly figurative, like many of Jesus's other parables, which He used to share concepts that were not easy to comprehend. Not only did Jesus refer to Himself as "the bread of life," but the Gospel of John records other declarations Jesus made about Himself as well, such as "the light of the world" (8:12), "the true vine" (15:1), "the good shepherd" (10:11), and His reference to His body as "the Temple" (2:9). In the last example, He certainly did not mean to imply He was an actual architectural structure!

Without a doubt, transubstantiation is not biblical, or even realistic. But by reading John 6 53–56 alone, one could possibly interpret it that way. However, after reading the entire chapter in context, it becomes unmistakably clear that Jesus only referred to sacrificing His body as spiritual food necessary for the redemption of our souls, not for consumption as physical food, like the manna from heaven. Obviously, when Jesus spoke the words "This is my body" and "This is my blood" during the Last Supper, He was only repeating His teaching of John 6, once more reminding His disciples that the bread and the wine He was blessing was only a symbolic representation of Himself, His body, His flesh, and His blood.

We must ask if the Catholic Church could possibly be wrong about their transubstantiation doctrine, and has the Church ever been wrong before? To arrive at an honest answer, we must seriously deliberate the reputation and the credibility of the Church. After a scrutinized review of its history, it is probably more appropriate to ask the question, has the Catholic Church ever been right? For most of its existence, it is sad to say, the Catholic Church has been an insult, disgrace, and an embarrassment to true Christianity. Catholic doctrine went through a gradual departure from New Testament truth. The Church deviated from what it supposed to be—an institution that could offer guidance and direction to the followers of Christ. The dominating factor of

Church leadership transformed into the pursuit of greed, power, and wealth. The Church hierarchy became the most corrupt and criminal organization on the face of the earth and completely lost regard for the Word of God. Popes declared themselves infallible, meaning that whatever they said would supersede the Word of God, demoting the Word of God to second place, so to speak. And it became a general practice for the Church to simply murder anyone who opposed, or even questioned, their arrogant behavior.

The Catholic Church reached its lowest point during the Middle Ages, known as the Dark Ages (500–1500). Europe was ruled by the popes, bishops, and priests as well as secular kings, and biblical Christianity was declared illegal, meaning that reading the Holy Bible, or the Word of God, was not permitted and was punishable by death. These corruptions resulted in repeated attempts by serious Christians to reform and perfect the doctrines of the Church, reintroducing Christianity back into the Church, so to speak. However, as the hope of reforming the Church from within diminished as a result of the Church's perceived arrogance, these reformers, protestors, or Protestants were simply tortured and murdered, usually by burning at the stake. A few survivors, for instance Martin Luther and Menno Simons, were forced to completely separate from the Church and establish their own Christian denominations. Shamefully, several Catholic historians still tend to deny this brutal and severe behavior and persist in defending the actions of the Church, arguing heresy rather than admitting that these people were true Christians and were doing nothing wrong; they just placed the Word of God before the doctrines of the Church.

Lasting for a period of about six centuries, the Inquisition was an ecclesiastical procedure of the Catholic Church established for the sole purpose of discovery and punishment of heresy. The Inquisition exercised immeasurable power and brutality in medieval and early modern times. The Inquisition's objective was to suppress the rights of all suspected heretics, depriving them of their estates and assets, which then became subject to the ownership of the Catholic treasury. Victims, predominantly innocent, were falsely accused, tortured into confession, and finally put to death. Witnesses for the defense were not allowed. The Inquisition became the legal framework throughout most of Europe that orchestrated one of the most outrageous religious orders in the course of humanity. While controversy rages around the number of victims of the Inquisition, conservative estimates easily place the count in the millions, although the Catholic Church admits only to somewhere between five thousand and seven thousand victims.

During His life here on Earth, did Jesus ever murder, or even despise, anyone who did not believe in Him or questioned His teaching? Why do some Catholic historians still insist that the murder of these innocent Christian believers were justified? Was the Church not aware of God's Ten Commandments, and was the Church not aware that murder was a capital sin? Or did they just conveniently bypass the Word of God? Which part of "Thou shall not bear false witness," "Thou shall not steal," and "Thou shall not murder" did they have difficulty understanding? The Church itself teaches that "human life is sacred because from its beginning it involves the creative action of God and it remains forever in a special relationship with the Creator, who is its sole end. God alone is the Lord of life from its beginning until its end: no one can under any circumstance claim for himself the right directly to destroy an innocent human being" (*Donum Vitae*, 5). Then, in direct contradiction to their teaching, the Church states that anyone who attempts to construe a personal view of *God* which conflicts with Church dogma must be burned without pity.

The Catholic Church also teaches that the souls of believers must burn in purgatory after death until all of their sins have been completely eradicated. To speed up this purging process, the Church proclaimed that bribes might be paid to the Church to persuade God to execute a premature release of certain souls. This heresy began creeping into the Catholic Church during the reign of Pope Gregory around the end of the sixth century, and it has no scriptural support. In fact, the Bible warned us about such pagan practice. Psalm 49:6–7 clearly tells us that a person could not redeem a loved one even if such a purgatory really did exist. "They that trust in their wealth, and boast themselves in the multitude of their riches; None *of them* can by any means redeem his brother, Nor give to God a ransom for him." The Catholic Church is clearly a human, not a holy, institution; it is not protected from sin. Even popes have been corrupt and sinners, many of them quite notorious.

Jesus was crucified together with two common criminals. Just prior to death, the one on the right humbly lifted his head and said, "Jesus, remember me when thou comest in thy kingdom." Then Jesus slowly lifted His head and looked at him, "and he said unto him, Verily I say unto thee, To-day shalt thou be with me in Paradise" (Luke 23:42–43). Note that Jesus literally said "today" and not "I will see you in Paradise as soon as you have done your time for your crime!"

At the beginning of the twenty-first century, Christianity had about 2.1 billion believers, meaning that about one out of every three people on the face of Earth was Christian, making Christianity the largest religion in the world. The largest Christian denomination is still the Catholic Church, with a membership of about 1.66 billion believers, suggesting that not all Christians are Catholic. However, sadly, not all Catholics are Christians either. Few Catholics believers remain strictly Catholic, meaning they have no consideration for anything other than the Church; they convince themselves that the Church is Holy, has never done anything wrong, and has never made a mistake, even though during the last few decades popes have practically traveled to the farthest reaches of the globe, apologizing for their mistakes and begging for forgiveness for the sins committed by the Church over the centuries. Strict Catholics seem to ignore this and still exalt the pope and the Church above God, still believing that the pope is infallible—that whatever the pope proclaims supersedes the Word of God.

When Jesus spoke the words "This is my body" and "This is my blood," He availed His body and His blood to be sacrificed for the redemption of all the sins of the world, including sins committed by the Roman Catholic Church. During the last few decades, the Catholic Church has obviously been attempting to become a more honorable and respectful Christian institution, and I believe that Christians must assist the Church in doing so. The Church is indeed in the process of making dramatic changes to its doctrine so it is more in accordance with what the reformers suggested in the first place. The Church is no longer a competitor in the political arena in Europe; most of the territories of the papal states were absorbed into the Kingdom of Italy in 1860 and reduced to a forty-four-acre sovereign city-state in the center of Rome. With a population of just over eight hundred, the Vatican is the smallest independent state in the world, both by area and population. Believers are now permitted to read and study the Bible, or the Word of God, without being persecuted, condemned, and murdered for doing so. Believers are no longer being charged with heresy, tortured, and murdered by burning for discussing or challenging the doctrines of the Church. The Church finally acknowledges the Ten Commandments; mass murder and genocide are no longer standard Church practice. Popes no longer bribe, torture, and murder their opponents, although in his book *Angels and Demons*, Dan Brown seems to suggest otherwise. Personally, I am truly delighted to see that slowly but surely the Catholic Church is transforming into a truly respectful honorable Christian institution. "Roma locuta est, causa finita est" (Rome has spoken, the case is closed) no longer applies in today's Christian society.

President John F. Kennedy was assassinated during an election campaign as he traveled in an open motorcade in Dallas, Texas, at 12:30 p.m., November 22, 1963; Texas governor John Connally was also injured. Within two hours, Lee Harvey Oswald was arrested for the murder of Dallas policeman J. D. Tippit and arraigned that evening. At 1:35 a.m. Saturday, Oswald was also arraigned for murdering the president. At 11:21 a.m., Sunday, November 24, 1963, the owner of the Carousel nightclub in Dallas, Jack Ruby, shot and killed Oswald as he was being transferred from Dallas police headquarters to the county jail. In 1964, the Warren Commission, investigating the assassination, concluded that there was no evidence that Oswald was involved in a conspiracy to assassinate the president and stated their belief that he had acted alone. Even before the publication of the official government conclusions, critics suggested a conspiracy was behind the assassination. Though the general public initially accepted the Warren Commission's conclusions, by 1966 the tide had turned, and authors (such as Mark Lane, with his best-selling book *Rush to Judgment*) and prominent publications such as the *New York Review of Books* and *Life* openly disputed the findings of the commission. One of the main concerns was the large number of suspicious or unexplained witness deaths connected with the investigation. Almost seventy witnesses, reporters, and law enforcement workers associated with the assassination died during two years after the Warren Commission investigation, most of them witnesses that insisted on telling the truth instead of submitting to the investigators falsified sequences of events. Investigator Jim Marrs points out that "these deaths certainly would have been convenient for anyone not wishing the truth of the JFK assassination to become public." We will probably never know for sure if these key witnesses were murdered to keep them silent or if they were genuine accidents or suicides. But there is no doubt these deaths have significantly hindered the official investigations and the public's understanding of what precisely took place at Dealey Plaza on November 22, 1963. The evidence that there was a cover-up is just as impressive as the evidence that there was a conspiracy in the first place.

James Henry Fetzer (b. 1940), an American philosopher, professor emeritus at the University of Minnesota in Duluth, and a well-known conspiracy theorist noted a series of inconsistencies with the Warren Commission's version of events, which he claims proves positively that its narrative is impossible, flawed, and untrue, and therefore is undeniably a cover-up. He claims that evidence released by the Assassination Records Review Board substantiates these concerns without a doubt. This evidence includes problems with bullet trajectories. Also, the alleged murder weapon, a 7.65 mm German Mauser rifle was found at the Texas School Book Depository, from where Oswald allegedly had fired the three shots that killed the president. When the FBI later reported that Oswald had purchased a 6.5 mm Italian Mannlicher-Carcano, the weapon at police headquarters in Dallas miraculously changed its caliber, its make, and its nationality. The Warren Commission suddenly concluded that a 6.5 mm Mannlicher-Carcano, not a 7.65 mm German Mauser, had been discovered by the Dallas deputies.

Here is a brief rundown of inconsistencies, including several points not mentioned in the text above:

- Oswald's description was broadcast over police radio within fifteen minutes of the assassination. There is no one who can explain how this description was obtained.
- No interrogation records were kept for those arrested at Dealey Plaza or for Oswald.
- The picture of Oswald holding a gun on the cover of *Time* magazine appears to be faked.
- JFK's body was removed from Dallas before an autopsy could be performed there.
- JFK's corpse left Dallas wrapped in a sheet inside an ornamental bronze casket. It arrived at Bethesda Naval Hospital in Washington in a body bag inside a plain casket; forensic evidence had been tampered with.
- The autopsy photographs of JFK's wounds differed radically from the descriptions given by the doctors at Parkland Hospital.
- A whole tray of evidence, including what was left of the president's brain, disappeared. It remains missing from the National Archives to this day, and nobody can explain where it went.
- The pristine condition of the "magic bullet" suggests it was planted.
- Numerous 8 mm home movies made by witnesses to the event were confiscated and never returned.
- Both the FBI and the CIA concealed important evidence from the Warren Commission.
- Critical data unearthed by the Warren Commission in 1964 and the Assassination Committee in 1978 are still classified as secret.
- Evidence of other gunman at the Grassy Knoll or the Dal-Tex Building, located across the street from the Texas School Book Depository, was briefly investigated but then ignored and not taken seriously.
- How did Jack Ruby find out about the exact details of Oswald's transfer, and how did he gain entrance into the highly secure Dallas police headquarters?

Fetzer maintains that John F. Kennedy was assassinated as the result of a well-planned and precisely executed conspiracy; even the shooters probably did not have any idea who gave the orders, who was involved, or who they were taking orders from.

Obviously the American people have been deceived and betrayed by the highest levels of their government, the very same people that have sworn to protect them.

A 2003 Gallup poll reported that 75 percent of Americans do not believe that Lee Harvey Oswald acted alone. That same year, an ABC News poll found that 70 percent of respondents suspected that the assassination involved more than one person. A 2004 Fox News poll found that 66 percent of Americans thought there had been a conspiracy, while 74 percent thought there had been a cover-up. The truth of the matter is that the truth will never be made public; the American people will never know exactly what happened.

On September 11, 2001, nineteen al-Qaeda terrorists hijacked four commercial passenger jet airliners. The hijackers intentionally crashed two of the airliners into the Twin Towers of the World Trade Center in New York City, killing everyone on board and many others working in the buildings. Both buildings collapsed within two hours, destroying at least two nearby buildings and damaging others. The hijackers crashed a third airliner into the Pentagon, and a fourth plane crashed in a field near Shanksville, Pennsylvania, after the passengers and flight crew revolted. The nineteen hijackers were all fanatic Muslim terrorists linked to the al-Qaeda terrorist organization and its leader, Osama bin Laden. Osama bin Laden was immediately identified as the perpetrator, and George W. Bush organized an intense international manhunt for bin Laden.

This is the official story. But what did really happen on September 11, 2001? Or, more importantly, what happened in the six months prior to September 11, 2001? From the very first day after 9/11, there were many incidents that puzzled many people. Soon it became quite apparent that the government had become engaged in a massive cover-up of what really did happen, including their own ties to the hijackers. A lot of unanswered questions still remain today regarding how the Twin Towers and the forty-seven-story 7 World Trade Center were brought down, and whether it was a Boeing 757 or a drone or a missile that hit the Pentagon, and if a hijacked airliner really did crash in Pennsylvania. Either the Bush administration was aware that these incidents were about to take place and they allowed them to proceed, or they were involved in the organization of the operation themselves—or perhaps there was a combination of both. The big, obvious question is why.

The Bush administration got off to a bad start. George W. Bush was the Republican candidate during the 2000 election; Al Gore was the Democrat. Al Gore was the actual new president elect; he got the most votes. However, George W. Bush ended up in the White House. Why? Because Bush cheated. George W. Bush won the 2000 presidential election by the narrowest margin in more than one hundred years. He lost the popular vote, meaning that more Americans voted for Al Gore than for him. In order to take charge of the Oval Office, a candidate must win 270 Electoral College votes. Bush got in with 271. The key state was Florida, whose 25 Electoral College votes decided the 2000 election. Controversy over the methods of vote counting, the handling of a recount, and the influence wielded by the Republican Party has raged ever since. In the official final tally, Bush won the state, which has a population of more than 18 million people, by just 537 votes despite losing the popular vote by 543,895.

George W. Bush, with the backing of the notoriously tough Republican Party, bribed, cheated, and swindled his way to the narrowest of victories in the sun-and-fun state. He did so by using the position of his brother Jeb as governor of Florida, and also through the help of Kathleen Harris, who operated Database Technologies, the maker of Florida's electronic voting system, which could block hundreds of thousands of democrats from voting or cause their votes not to be counted. The scandal over the voting irregularities ended up in the Florida Supreme Court, and eventually the US Supreme Court ruled in Bush's favor, likely because the judges on the Supreme Court had been appointed by Bush's daddy, then President George H. W. Bush. It was James A. Baker III—senior partner in the high-profile law firm

Baker-Botts of Houston, Texas, and secretary of state under the first Bush administration—who directed George W. Bush's legal affairs with respect to the contested 2000 presidential vote in Florida. It was Baker's strategy to have the election decided in the US Supreme Court instead of through more traditional means, such as counting the ballots. "The vote in Florida has been counted," Baker said, "and the vote in Florida has been recounted. Governor George W. Bush was the winner of the vote, and he was also the winner of the recount. Based on these results, we urge the Gore campaign to accept the finality of this election."

Obviously the American people voted for a Democrat but ended up with a Republican president they did not want. George W. Bush did not win the presidency; George W. Bush actually stole the presidency. For the first time ever in the history of US politics, a president was appointed to the Oval Office by the US Supreme Court rather than being elected.

President George W. Bush's inauguration took place on January 20, 2001. He was the first and only president ever to arrive at the White House amid protesting and demonstrating crowds, being called a cheater; his limousine was hit by a tennis ball and an egg. During Bush's first eight and a half months in the Oval Office, things did not get any better and remained rather unusual. He lost Republican control of the Senate, and according to the *Washington Post*, he spent 42 percent of his time on vacation.

Doing business with the enemy is nothing new to the Bush dynasty. The Bushes have always been comfortable putting profits before patriotism. Michelle Mairesse wrote in her essay *The Bush-Saudi Connection*, "The Bushes are carriers of the deny-destroy-and-be-damned virus. Prescott Bush never apologized for trading with the Nazis. George Bush Senior professed to know nothing of the drug and arms dealing operations, although it was common knowledge that he directed them. He and his sons enriched themselves through shady real estate deals and financial manipulations that brought down entire banking and savings and loan institutions. They are all consummate inside traders, looting and leaving ruin in their wake."

Shortly after the 9/11 attack, Osama bin Laden was immediately identified as the perpetrator, but not many people are aware of the fact that the Bushes and the bin Laden's were not exactly strangers. Osama's eldest brother, Salem, was one of President George W. Bush's first business partners. George W Bush and the bin Laden family have been connected through dubious business deals since 1977, when Salem, the head of the bin Laden family business, one of the biggest construction companies in the world, invested in Bush's start-up oil company, Arbusto Energy Inc. James R. Bath, Bush's friend, neighbor, and fellow pilot in the US National Guard, was involved to funnel money from Osama bin Laden's brother, Salem, to initiate George W. Bush in the oil business. Salem bin Laden, a close friend of the Saudi king Fahd, had "invested heavily in Bush's first business venture," according to the *Daily Mail* (UK). However, the company faced financial collapse and bankruptcy in September 1984; it was then merged with Spectrum 7, an energy company, in an effort to keep it afloat. As the hard times continued, Spectrum 7 merged with Harken Energy in 1986. Two months prior to the Iraqi invasion of Kuwait, on June 20, 1990, George W. Bush sold two-thirds of his Harken stock—212,140 shares at $4 a share, for a total of $848,560. Eight days later, Harken finished the second quarter reporting losses of $23.2 million and the stock went into a nosedive, losing 75 percent of its value, finally settling at a price of a little over $1 a share. Although the law requires prompt disclosure of insider sales or sales by senior executives, Mr. Bush did not report the incident to the Securities and Exchange Commission (SEC) until 34 weeks later. The SEC investigated, but it was the prestigious law firm Baker Botts of Houston that was instrumental in getting Bush off the hook.

The Carlyle Group, located on Pennsylvania Avenue in Washington, DC, midway between the White House and the Capitol, was founded in 1987 by David Rubenstein, a policy assistant in Jimmy Carter's administration, and two of his lawyer friends. The Carlyle Group is an investment firm managing some $97.7 billion in assets, including interest in a number of defense-related companies. Carlyle has been successful in attracting former politicians to its staff, including former president George Bush Sr. and his secretary of state, James A Baker III; former British prime minister John Major; one-time World Bank treasurer Afsaneh Masheyekhi; several Southeast Asian powerbrokers, former German Bundesbank president Karl Otto Pohl; and former chairman of the SEC, the US stock market regulator Arthur Levitt. Carlyle partners include James A Baker III, who is senior counselor for the Carlyle Group, and the firm's chairman, Frank Carlucci, who served as defense secretary in the Ronald Reagan administration and as deputy director of the CIA. Among the firm's multimillion-dollar investors were members of bin Laden's family. President George Bush Sr. visited the bin Laden's in Saudi Arabia on several occasions on behalf of Carlyle Group. However, in October 2001, after the 9/11 attacks, the bin Laden's withdrew their investments in the Carlyle Group because of continual criticism and conflict of interest. Judicial Watch, a public interest law firm that fights government corruption, says the connection between the Pentagon and the Carlyle Group, with the first president Bush as an advisor, creates the "appearance of conflict" and violates the public's trust.

It is interesting to note that on September 11, 2001, at the annual conference for the Carlyle Group at the Ritz-Carlton Hotel in Washington, DC, both George Bush Sr. and James A. Baker III were watching television coverage of the 9/11 attacks of the World Trade Center and the Pentagon together with Shafig bin Laden, who is the half-brother of Osama, who was accused of being the ringleader of the 9/11 attacks by the Bush administration. They were most likely attending the conference to explore the potential profits for United Defense from another pending war with Iraq. At approximately the same time, just after 9:00 a.m., President Bush arrived at Booker Elementary School in Sarasota, Florida, for a photo opportunity. President Bush was listening to and participating with second-graders taking turns reading a story called "The Pet Goat," a fairy tale about a little girl's pet goat. Suddenly Senior Advisor and Deputy Chief of Staff Karl Rove rushed up, took President Bush aside in a corridor, and informed him that an airliner had just crashed into the North Tower of the World Trade Center Complex. Rove also informed the president that the cause of the crash was unclear. President Bush replied casually, "What a horrible accident!" At 9:03 a.m., a second airplane hit the South Tower of the World Trade Center, and Chief of Staff Andrew Card approached the president and whispered in his ear, "A second plane hit the second tower, America is under attack." However, Bush did not panic. Instead he just got a strange look on his face and stared at a spot on the wall. He did not leave the classroom to obtain a further briefing or gather more information or find out precisely what was happening. Rather, he just sat there very relaxed, with a composed stare on his face, and continued to listen as the sixteen Booker Elementary School second-grade students continued to read the pet goat story. President Bush then picked up a copy of the book and read along with the children for about seven more minutes.

It is almost certain that the entire population of the United States, or perhaps the whole world, have their reservations about exactly what went through Bush's mind during those seven critical minutes. What was he thinking? Was he not concerned about the American people, and did he not realize or even care that America was being at attacked by some unknown force? His peculiar behavior certainly has the tendency to make one believe that he was he well aware of what was happening in New York and thought, *Oh my God, they did it; it is happening!* Or perhaps he just did not care about the safety and well-being of American citizens and found the story "The Pet Goat" much more exiting and interesting!

George W. Bush selected the following officials for his key staff members:

- Dick Cheney, who served two full terms as vice president (2001–2009)
- Colin Powell, as secretary of state (replaced by Condoleezza Rice after his resignation in 2005)
- Donald Rumsfeld, as secretary of defense (until 2006, followed by Robert Gates)
- John Ashcroft, as attorney general (until 2005, replaced by Alberto Gonzales until 2007, who was then followed by Michael Mukasey)
- Tom Ridge, as secretary of homeland security (until 2005, replaced by Michael Chertoff)
- Richard Clarke, who served as special advisor to the president

- Paul Wolfowitz, who was deputy secretary of defense from 2001 till 2005, after which he became the head of the World Bank.

Richard Alan Clarke (b. 1950) was a US government official for thirty years, from 1973 until his resignation in 2003. He worked for the State Department during the presidency of Ronald Reagan, and in 1992, President George H. W. Bush appointed him to chair the Counterterrorism Security Group and to a seat on the US National Security Council. President Bill Clinton retained Clarke and in 1998 promoted him to be the national coordinator for security, infrastructure protection, and counterterrorism as the chief counterterrorism adviser on the National Security Council. Under President George W. Bush, Clarke initially continued in the same position, but the position was no longer given cabinet-level access. He later became the special advisor to the president on cyber security before leaving the Bush administration in 2003.

On March 22, 2004, Richard Clarke's book *Against All Enemies: Inside America's War on Terror—What Really Happened* was published. The book was critical of past and present presidential administrations for the way they handled the War on Terror both before and after September 11, 2001. He focused much of its criticism on Bush for failing to take sufficient action to protect the United States from threats prior to the September 11, 2001, attacks and for the 2003 invasion of Iraq, which Clarke feels greatly hampered the War on Terror and was a distraction from the real terrorists.

Clarke claims President Bush made Iraq the "central front in the war on terror" and said that before and during the 9/11 crisis, the Bush administration were distracted from efforts against Osama bin Laden's al-Qaeda organization by a preoccupation with Iraq and Saddam Hussein. He also said that on September 12, 2001, President Bush pulled him and a couple of aides aside and "testily" asked him to try to find evidence that Saddam was connected to the terrorist attacks. In response he submitted a report stating there was no evidence of Iraqi involvement and got his report endorsed by relevant agencies, including the FBI and the CIA. The paper was quickly returned by a deputy with a note saying, "Please update and resubmit."

On August 6, 2001, President Bush received one of his few briefings on terrorism, titled "Bin Laden Determined to Strike in US." It would be the last such briefing he would receive. Even though Clarke was panicked about the potential attacks, President Bush told Clarke that he didn't want to be briefed on this again.

Finally the first Principal's Committee meeting took place on September 4, 2001, almost a week prior to the 9/11 attacks. This meeting had been requested and insisted upon by Clarke since January 25, 2001! At the meeting everyone seemed to agree with Clarke and FBI director George Tenet regarding the threat posed by al Qaeda. However, Donald Rumsfeld diverted attention from al Qaeda and insisted that there were other, more important, terrorist concerns, such as Iraq. In the early morning hours of September 12, the day after the attacks, Clarke walked into a White House meeting expecting to discuss the attacks; instead, he "walked into a series of discussions about Iraq."

Bob Woodward's books *Bush at War* and *Plan of Attack* assert that the Bush administration started planning for the war on Iraq immediately after taking office.

Former FBI deputy director John O'Neill was heavily involved in al Qaeda investigations but was prevented from investigating by George W. Bush after the president signed directive W199i, which made it a crime to "investigate or hinder the terrorists in any way." Out of frustration, O'Neill resigned his position with the FBI and was offered a position as director of security at the World Trade Center, at three and a half times his FBI salary, by Marvin Bush, the president's brother. Tragically, John O'Neil died during the 9/11 attack while helping people escape from the twenty-fourth floor of the first building.

Obviously, the Bush administration had absolutely no interest in al Qaeda or Osama bin Laden; their entire focus was directed to Iraq and Saddam Hussein. After 9/11, Osama bin Laden was supposed to be hiding out somewhere in a cave in Afghanistan; however, Donald Rumsfeld alleged there were no decent targets in Afghanistan and that Iraq had better targets. But why Iraq? Asked why a nuclear power such as

North Korea was being treated differently from Iraq, where no weapons of mass destruction had been found, Deputy Secretary of Defense Paul Wolfowitz commented to look at the situation simply, the most important difference between North Korea and Iraq is economically, they just had no choice in Iraq. The country swims on a sea of oil.

However, after 9/11 George W. Bush did organize an intense international manhunt for Osama bin Laden, guaranteeing the American public that bin Laden would be apprehended, captured, and brought to justice. Bush said on several occasions, "We will smoke him out of his hole!" But only six months later, on March 13, 2002, Bush apparently had a change of mind when he said, "I don't know where bin Laden is. I have no idea and really don't care. It's not that important. It's not our priority ... I am truly not that concerned about him"

It is interesting to note both that terrorists almost always claim responsibility shortly after their evil deeds and that Osama bin Laden denied any involvement in 9/11. Not many people are aware that, after September 11, 2001, Osama bin Laden issued a statement on tape that he had nothing to do with the attacks on America and that such actions were against the teachings of Islam. The American public, however, was denied access to this information; they were told instead that bin Laden could possibly have an embedded a secret code on the tape that would alert other terrorist sleeper cells to activate and target other American cities. On his tape Osama bin Laden said, "I was not involved in the September 11 attacks in the United States, nor did I have knowledge of the attacks. There exists government within a government within the United States. The United States should try to trace the perpetrators of these attacks within itself; to the people who want to make the present century a century of conflict between Islam and Christianity. That secret government must be asked as to who carried out the attacks ... The American system is totally in control of the Jews, whose first priority is Israel, not the United States."

Immediately after the 9/11 attacks, the Bush administration allowed Osama bin Laden's family members to leave the United States, even though the entire American airspace was closed to all air traffic. Richard Clarke acknowledged that the Bush administration permitted about

140 high-ranking Saudi Arabians, including twenty-four members of the bin Laden family, to depart from the United States, despite the fact that Secretary of State Colin Powell announced on September 13, 2001, that Osama bin Laden was the prime suspect in the 9/11 attacks. Interestingly, the FBI never detained or conducted proper interviews with bin Laden's family members before allowing them to flee the country. The Bush administration initially denied these flights ever took place, but in 2003, National Security Agency official Richard Clarke and Secretary of State Colin Powell confirmed these flights actually did take place.

The World Trade Center towers were not exactly the real-estate marvels everybody thought they were. Initially intended to be a complex dedicated to companies and organizations directly involved in world trade, they at first failed to attract the expected clientele. During the early years, various governmental organizations became key tenants, including the State of New York. It was not until the 1980s that the city's risky financial state eased, after which an increasing number of private companies—mostly financial firms tied to Wall Street—became tenants. During the 1990s, approximately five hundred companies had offices in the complex, including financial organizations such as Morgan Stanley, Aon Corporation, Salomon Brothers, and the Port Authority of New York and New Jersey itself. From an economic point of view, the World Trade Center was not doing very well and had been subsidized since its start by the Port Authority of New York and New Jersey.

Up until the early seventies, asbestos fibers were added to flame-retardant sprays used to insulate steel building materials, particularly floor supports. The insulation was intended to delay the steel from melting in the case of fire by up to four hours. After asbestos workers became ill from high exposures to the fibers, the use of asbestos as spray fireproofing was banned by New York City In 1971. However, at that time, asbestos insulating material had already been installed up to the sixty-fourth floor of the World Trade Center towers.

The Twin Towers required some $200 million in renovations and improvements, mostly related to the removal and replacement of these asbestos building materials. For years, the port authority treated the buildings like aging white elephants, attempting on several occasions to get permits to demolish the buildings for liability reasons, but the permits were denied because of the known asbestos problem. The only reason the buildings were still standing on September 11, 2001, was because it was too costly to manually disassemble the towers floor by floor. The projected cost to disassemble the towers was an estimated $15 billion. The scaffolding alone for the operation was estimated at $2.4 billion!

Larry Silverstein, a generous contributor to Democratic and Republican office-holders, is a New York property tycoon who purchased the entire World Trade Center Complex, all seven buildings, just six months prior to the 9/11 attacks. That was the first time in its thirty-three-year history the complex had ever changed ownership. Mr. Silverstein's first order of business as the new owner was to change the company responsible for the security of the complex. The new security company he hired was Securacom, later renamed Stratasec. The brother of President George W. Bush, Marvin, was on its board of directors, and Marvin's cousin, Wirt Walker III, was its CEO. According to public records, not only did Securacom provide electronic security for the World Trade Center "up to the day the buildings came down," but it also covered Dulles International Airport, from which one of the hijacked airplanes departed, as well as United Airlines, another major player in the 9/11 attacks.

A number of banks and investors provided most of the funds for the deal. Larry Silverstein, president of Silverstein Properties, made a down payment of $124 million on this $3.2 billion purchase, only using $14 million of his own money. Then he promptly arranged for a lucrative insurance policy, insisting that the entire complex be covered against terrorist attacks.

An important aspect of 9/11, hardly ever mentioned in the media and also omitted from the 9/11 Commission Report, was the demolition of 7 WTC about seven hours after the Twin Towers collapsed. According to the government's vague and inconclusive reports, fires caused 7 WTC to collapse. Yet, except 9/11, there has never been a case where fire, no matter how severe, caused the collapse of a steel-framed high-rise building. In the case of 7 WTC, the building was not hit by an airplane or significant debris from the collapsing tower some three hundred feet away, and with just a few minor fires burning inside, at 5:20 p.m. this massive concrete-and-steel-framed forty-seven-story skyscraper imploded into its own footprint in less than seven seconds. Without any doubt, the collapse of 7 WTC was a controlled demolition. It is simply not possible to make a decision to "pull" a building and have it collapse like it did just a few moments after. In fact, implosions take a minimum of two weeks to several months to rig and prepare, and these preparations are performed by highly trained, experienced professionals who plan and test far in advance. Obviously the preparation for demolition, or rigging the building for demolition, was already completed well in advance of the 9/11 attack and therefore indicates a conspiracy. The owners had already planned to demolish not only the Twin Towers, but also 7 WTC, well before 9/11. Was September 11 just their lucky day, or was it a well-planned and coordinated event? I am sure the truth is out there somewhere, but the general public will never know.

Larry Silverstein, the new owner of the building, sort of let the cat out of the bag, admitting to a controlled demolition, when he made the following statement: "I remember getting a call from the, er, fire department commander, telling me that they were not sure they were gonna be able to contain the fire, and I said, 'We've had such terrible loss of life; maybe the smartest thing to do is pull it.' And they made that decision to pull, and we watched the building collapse" (PBS, September 10, 2002). After consultation, the fire department and Silverstein made the decision to pull, or demolish, the building, and in fact, just a few moments after the decision, the building collapsed and came down in less than seven seconds. This particular incident probably holds the key to unlocking the entire 9/11 mystery.

Mr. Silverstein told journalist Steven Brill that he was so sickened by the destruction on 9/11 and the deaths of four of his employees that he did not focus on insurance or financial matters until "perhaps two weeks later." But according to two people who called him on the morning of 9/11 to offer their sympathy, Silverstein soon changed the subject: "He had talked to his lawyers … and he had a clear legal strategy mapped out. They were going to prove, that the way his insurance policies were written the two planes crashing into the two towers had been two different 'occurrences,' not part of the same event. That would give him more than $7 billion to rebuild, instead of the $3.55 billion that his insurance policy said was the maximum for one 'occurrence.' And rebuild was just what he was going to do, he vowed." By midmorning, he called his architect David Childs and instructed him to start sketching out a plan for a new building. He told Childs to plan to build the exact same area of office space as had been destroyed. Following a lengthy legal dispute, Silverstein would eventually receive $4.55 billion in insurance payouts for the destruction of the WTC.

The World Trade Center controlled-demolition conspiracy theory suggests that the collapse of the World Trade Center was not caused by the plane-crash damage that occurred as part of the September 11, 2001 attacks, nor by resulting fire damage, but by explosives installed in the buildings in advance. Demolition theory proponents, such as physicist Steven E. Jones, architect Richard Gage, software engineer Jim Hoffman, and theologian David Ray Griffin, argue that the aircraft impacts and resulting fires could not have weakened the buildings sufficiently to initiate a catastrophic collapse, and that the buildings would have collapsed neither completely nor at the speeds that they did without additional energy involved to weaken their structures. Jones has suggested that thermite or super-thermite was used to demolish the buildings and says he has found evidence of such explosives in the WTC debris.

Dick Cheney was George H. W. Bush's defense secretary and served under George W. Bush as vice-president. After his term in office as defense secretary under George H. W. Bush, Cheney became chairperson and CEO of the world's second-largest oil service company, Halliburton, in which position he served from 1995 to 2000. Cheney retired from Halliburton during the 2000 US presidential election campaign with a severance package worth $36 million.

But why was Dick Cheney so eager to invade Iraq? Why did he repeatedly link Saddam Hussein to al Qaeda after September 11, and why did he maintain not only that Iraq had weapons of mass destruction but that he, Cheney, knew exactly where they were? Cheney clearly came into office insisting on a war with Iraq, as mentioned on several occasions by former US Treasury secretary Paul O'Neill. Even prior to the invasion of Iraq, without a tender, Halliburton was already awarded contracts worth up to $7 billion to repair the Iraqi oilfields. Is it not likely that Mr. Cheney's position as CEO at Halliburton, and his $36 million severance package had something to do with these unbidden contracts, while he still continued to receive deferred income from Halliburton? Is this the reason that the war on terror was diverted from al Qaeda to Iraq?

Halliburton's business with the federal government has grown significantly since the Bush administration took office. As the second-largest oil service and engineering company in the world, it was given a direct line to the White House. According to the *New York Times*, Halliburton went from being the twenty-second-biggest military contractor in 2000 to the seventh-largest in 2003. It was also given contracts to provide logistical support to US troops, handling everything from food to transport and laundry services. That arrangement, awarded under an existing long-term contract to provide emergency services, was worth a potential $13 billion. When the first secretive contract was exposed, shortly after the March 2003 invasion, Democrats argued there was evidence of a loyalist network in the White House.

According to a June 1, 2004, Reuters article, there were "fresh calls on Capitol Hill" for investigations into whether Dick Cheney was instrumental in Halliburton getting these lucrative deals. A March 2003 Pentagon e-mail said an unbid Halliburton contract to rebuild Iraq's oil industry was "coordinated" with Cheney's office. The e-mail, reported in the June 7, 2004, issue of *Time* magazine, provided "clear evidence" of a relationship between Cheney and multibillion-dollar Halliburton contracts. Senator Patrick Leahy commented, "It totally contradicts the vice president's previous assertions of having no contact with federal officials about Halliburton's Iraq deals." Leahy, a Vermont Democrat, made this statement in a conference call set up by John Kerry's presidential campaign.

Yet Bush relentlessly led the United States into a war that has cost the lives of 4,747 (as of June 2010) young American soldiers during Operation Iraqi Freedom, and 2,242 young Americans died in Afghanistan during Operation Enduring Freedom Afghanistan. More than 33,000 were seriously injured. Did all these young Americans die in honor of their country? No, they certainly did not. The war with Iraq was based solely on the presence of weapons of mass destruction, which the Bush administration knew did not exist. The Bush administration went to war with Iraq under false pretenses based on a pack of lies. These young Americans were led to their deaths based on the lies of a corrupt administration. They did not die in defense of their country; sadly, the simple cause for their demise was to make very few people very rich. During the conflict, 150,726 civilian and combatant Iraqis have also been killed. It is interesting to note that Iraq is a nation that has never invaded the United States, nor has it ever threatened to invade the United States. In fact, Iraq had never even killed one single American civilian. Where I really got sick to my stomach was when I was listening to Donald Rumsfeld in a White House briefing explaining how much "care and humanity" went into to blowing the Iraqis away.

According to the Center for Defense Information, the estimated cost of the wars in Iraq and Afghanistan reached $1.08 trillion by the end of fiscal year 2010. President Bush betrayed the trust of the American people; all those weapons of mass destruction turned out to be nothing more than a lot of words of mass deception. It was like the old Dutch proverb, "Even if the lie is fast and kind, the truth, is never far behind!"

Shortly after the Iraq invasion, the Central Intelligence Agency, Defense Intelligence Agency, and other intelligence agencies also discredited evidence related to Iraqi weapons and links to al Qaeda. At this point the Bush and Blair administrations began to shift to secondary justifications for the war, such as Saddam Hussein's government's human rights record and the promotion of democracy in Iraq.

In April 2007, Representative Dennis Kucinich, an Ohio democrat, filed an impeachment resolution (H.R. Res. 333) against Vice President

Dick Cheney, seeking his trial in the Senate. The resolution charged that Vice President Cheney

1) had purposely manipulated the intelligence process to deceive the citizens and Congress of the United States by fabricating a threat of Iraqi weapons of mass destruction,
2) had fabricated a threat about an alleged relationship between Iraq and al Qaeda in order to justify the use of the US Armed Forces against Iraq in a manner damaging to US national security interests, in violation of his constitutional oath and duty, and
3) had openly threatened aggression against Iran absent any real threat to the United States, and had done so with the proven US capability to carry out such threats, thus undermining US national security.

After months of inaction, Kucinich reintroduced the exact content of H.R. Res. 333 as a new resolution numbered H.R. Res. 799 in November 2007. Both resolutions were referred to the Judiciary Committee immediately after their introduction, and the Judiciary Committee did not consider either. Both resolutions expired upon the termination of the 110th United States Congress on January 3, 2009.

The airplane that flew into the WTC North Tower was reported by the Bush administration to be American Airlines Flight 11, a Boeing 767, registration number N334AA, carrying ninety-two people, including five Arabs who had hijacked the plane.

The second aircraft, which crashed into the South Tower, also a Boeing 767, was identified as a United Airlines flight, registry number N612UA, carrying sixty-five people, including the crew and five hijackers.

American Airlines Flight 77 was reported to be a Boeing 757, registration number N644AA, carrying sixty-four people, including the flight crew and five hijackers. This aircraft, with a 125-foot wingspan, was reported to have crashed into the Pentagon, leaving an entry hole no more than sixty-five feet wide.

United Airlines Flight 93 was reported by the federal government also to be a Boeing 757 aircraft, registration number N591UA, carrying forty-five persons, including four Arab hijackers who had taken control of the aircraft. For some unknown reason, the plane crashed into a farm field in Pennsylvania.

Every passenger-carrying aircraft consists of thousands of critical parts, most of which are virtually indestructible and individually identifiable through serial numbers. When these parts are installed, their serial numbers are coordinated with the aircraft's serial number in the records of the aircraft. After a certain number of flying hours, these aircraft are required to undergo crucial inspections, performed by specialist aeronautical mechanics, and some of these parts are required to be replaced or overhauled. Most of these parts, whether they are hydraulic pumps, flight surface actuators, landing gears, engines, or engine components, are virtually indestructible either through impact or fire. It would be impossible for a fire resulting from an airplane crash to destroy or obliterate these critical parts with their unique serial numbers. However, the US government has never produced any identifiable parts, which would prove beyond any doubt that these aircraft actually crashed at their specific sites. The only evidence that two large aircraft, the size of a Boeing 767, hit the North and the South Towers of the World Trade Center is the video footage shot by amateur and professional photographers. From this footage, the airplanes themselves are impossible to positively identify. It seems that all potential evidence was deliberately kept hidden from the public or destroyed. These facts surely suggest that the 9/11 hijackings were part of a black operation carried out with the cooperation of elements within the US government.

The Pentagon attack has probably attracted more conspiracy-related debate than any other 9/11 issue. Hani Hanjour, one of the nineteen presumed Arab terrorists, is supposed to have taken over the controls of Flight 77 and piloted the Boeing 757 into the Pentagon. However, Hanjour was known in aviation circles as an incompetent pilot not even capable of mastering the controls of a light Cessna 172 aircraft. How could such an inexperienced Mickey Mouse pilot, who had never even touched the controls of a Boeing 757 before, possibly make a complicated 270° turn while descending from seven thousand feet and hit a specific section of the Pentagon as alleged? This would be quite a challenge for even the most experienced jet fighter pilot; for Hanjour it would be virtually impossible! With all the evidence readily available at the Pentagon crash site, any unbiased, logical investigator could only conclude that it was not a Boeing 757 that crashed into the Pentagon as alleged.

What about the fiasco regarding the identification of the alleged hijackers? The full list of the nineteen hijackers' names was first released on September 14, 2001. However, shortly thereafter seven of these men were confirmed as being alive and well and having no connection to 9/11.

On September 17, 2001, the British newspaper the *Independent* reported that one of the alleged "suicide hijackers" was actually an airline pilot living in Jeddah, Saudi Arabia. The *Independent* reported hat Abdurrahman al-Omari, a pilot with Saudi Airlines and was astonished to find himself accused of being a hijacker as well as being dead and has subsequently visited the US consulate in Jeddah to demand a full explanation.

A second alleged hijacker named Saeed Al-Ghamdi was questioned shortly after the attacks and faced further humiliation when CNN flashed a photograph of him around the world.

Then a third alleged hijacker, Khalid Al Midhar, was also found to be alive and well. FBI Director Robert Mueller at this point acknowledged that the identities of several of these perpetrators were now questionable, while Saudi Airlines was threatening legal action against the FBI for putting its reputation in jeopardy.

Saudi Airlines pilot Saeed Al-Ghamdi and Abdulaziz Al-Omari, an engineer from Riyadh, are furious that the hijackers' 'personal details'— including name, place, date of birth and occupation—matched their own. Yet the incredible disclosures of alleged hijackers turning up alive continued.

Waleed al-Shehri was identified by the FBI as one of the Flight 175 hijackers, but he turned up alive and well in Casablanca, Morocco, on September 23, 2001. He said he was a pilot with a Saudi Arabian airline and was training in Morocco. Al-Shehri identified himself after his name and photograph were published in the world media, but he said he had nothing to do with the attacks and was a victim of misidentification.

But perhaps the most fascinating form of "proof" was the miraculous discovery of hijacker Satam Al-Suqami's passport a few blocks away from the crash site. The World Trade Center fires were fierce enough, we are told, to melt steel and destroy both virtually indestructible black boxes from the airplanes. Yet a flimsy passport from one of the terrorists survived the inferno and landed gently on a side street. None of these hijackers' names were shown on the passenger manifests of the hijacked flights, suggesting they never even boarded these flights.

What actually did 9/11 accomplish? The 9/11 attack accomplished a great many things. First of all, it took care of all the problems with the Twin Towers. These buildings had become a thorn in the side for the owners, the Port Authority of New York and New Jersey. The asbestos dilemma was too expensive to fix, and a multitude of expensive lawsuits were anticipated. The Twin Towers represented a serious problem for which there was no simple answer. The owners were, as the saying goes, "caught between a rock and a hard place." How convenient that 9/11 made all the port authority's problems instantly disappear. And where did Larry Silverstein come from? Suddenly he purchased the entire World Trade Center Complex, being very well aware of all the problems associated with the buildings, but he still goes ahead and buys them anyway. He then buys a very lucrative insurance policy, including unusual protection against terrorist attacks. Then, just a few weeks later, the WTC falls victim to a terrorist attack, making Silverstein an instant billionaire. Wow!

The September 11 attacks paved the way for the Bush administration to launch the war with Iraq and rolled out the red carpet for the US Army to invade. Why? The answer to this question is very simple; the Bush administration used 9/11 to scare the living daylights out of the

American people. According to the administration, there could be a terrorist hidden in every corner of the United States, armed with a WMD. The administration even resorted to a color-coded threat warning system, like one would find in a national forest to remind visitors of the dangers of forest fires, except these were designed to indicate the danger of a terrorist attack. Of course, the arrow was always pointed to the red or the orange, never to the green or the blue. However, it came to be viewed by many as a complete joke and proved to be quite an embarrassment to the Bush administration.

Prior to George Bush Jr. taking office, Richard Clarke and FBI Special Agent John O'Neill were investigating allegations that al Qaeda was planning an attack within the United States and were trying very hard to bring the matter to the attention of the president and the administration. They knew that there was something brewing, that something bad was going to happen, but they were continually reminded that the administration did not want to hear about al Qaeda; they just wanted to hear about Iraq.

Did George Bush have anything to do with the whole affair? Most likely not, because most of his time was spent away from the office during the time 9/11 was developing.

It has become quite obvious that the Twin Towers and 7 WTC were rigged for controlled demolition well prior to 9/11. Everybody involved in the planning of 9/11 was well aware of the fact that just flying a large aircraft into the buildings would indeed cause considerable damage but would not demolish the buildings. It was common knowledge that the buildings had been designed to withstand the impact of a Boeing 707, the largest passenger airliner popular during the 1960s. So what happened? Was it a coordinated combined effort to make 9/11 a reality? Was this a joint venture, so to speak?

After the invasion of Iraq, it did not take the US Army very long to locate and apprehend Saddam Hussein—to "smoke him out of his hole"—and execute him shortly thereafter. The execution of Saddam Hussein took place on December 30, 2006. He was sentenced to death by hanging after being found guilty and convicted of crimes against humanity by the Iraqi Special Tribunal for the murder of 148 Iraqi

Shiites in the town of Dujail in 1982, in retaliation for an assassination attempt against him. Did George Bush not condemn 152 of 153 death row prisoners while he was governor of Texas and then laugh about it?

The Bush administration was well aware of the fact that Saddam Hussein had no connection to 9/11. Did the Bush administration use the 9/11 incident as an excuse to hang Saddam Hussein? How about Osama bin Laden? Well, he is probably still in hiding, however not in a dirty old cave somewhere in Afghanistan; more likely he is on a ranch somewhere in Texas owned by the Bushes, smoking big Havana cigars and living high on the hog.

There are probably very few people that actually know the real truth about 9/11, and I am sure they'll never tell, meaning common people will never know what really happened. The only thing we can do is seriously consider the facts we know are true and then draw our own conclusions and make up our own minds. The whole 9/11 incident is still surrounded in a shroud of suspicion and represents one of the darkest eras in the history of the United States.

It was not only the three thousand deaths on September 11, 2001, that were made possible by the above corrupt and criminal acts. The culture that enabled the 9/11 events to occur then plunged the United States into two wars in which the deaths and brutal injuries exceeded one million. It also created world-record numbers of people wanting to kill Americans and brought about an estimated $5 trillion in costs that are already impoverishing many people in the United States.

The Grand Finale

In the previous chapters we have examined the origin and finale of our universe and have logically concluded that there is not one person on the face of Earth that has a precise idea how the universe came into existence. Science has produced a multitude of theories and fairy tales, and the only evidence the Christian Bible reveals is that the universe was created by the word of God, although it does not provide explicit details or explanations precisely of how the universe came to be. However, where the Bible is more precise and detailed is in its description of how everything will come to an end. In the chapter "The End of Times," we compared Bible prophecy to a puzzle—the prophetic puzzle, so to speak. However, the last clue to this puzzle was reserved for the last chapter. This last, but most significant, clue is found in Matthew 24:14"And the gospel of the kingdom shall be preached in the whole world for a testimony unto all nations; then shall the end come." What this clearly indicates is that prior to the Rapture, or the Second Coming of the Lord Jesus, the Word of God will be preached to every tongue and nation on the face of Earth, and "then shall the end come." Today there are more than 6 billion copies of the Bible in circulation; the Bible is still the best-selling book of all time, with more than 78.5 million copies sold every year in more than 1,200 languages. The Bible is proclaimed in books, newspapers, magazines, television, over the radio, and via the Internet, globally. How long before the Bible will be proclaimed to every soul on Earth? Nobody can tell; nobody can possibly calculate the precise date. I do believe, however, that when this key event becomes a reality, the Rapture will occur, and this truly could happen at any moment. How long must we wait? Well, nobody knows. Matthew 24:36 provides the final clue: "But of that day and hour knoweth no one, not even the angels of heaven, neither the Son, but the Father only."

Be Prepared

As we approach the end of times, Jesus encourages us to be alert and guard our spiritual positions as well as world events and conditions. A time of unparalleled global disorder is also the threshold of the Lord's Second Coming. Believers today face the same dilemma as Joseph did in ancient Egypt and as Noah did before the flood. Just as Joseph and Noah were warned of global catastrophes, we have also been warned of devastating events to come. In the world today, and especially in America and Australia, there are millions of "preppers," people that are feverishly preparing for the end times. In 2012, National Geographic aired the reality TV show *Doomsday Preppers*. Since the broadcast of the documentary, the prepper phenomenon has taken on an added dimension. The show featured preppers from all walks of life and backgrounds who were engaging in a variety of extremes, from hoarding food to acquiring guns and ammunition to building bunkers. Most of the show's participants admitted that their religious convictions played an important role in their preparedness actions. What is a prepper? According to www.Prepper.org a prepper is described as an individual or group that prepares or makes preparations in advance of, or prior to, any change in normal circumstances or lifestyle without significant reliance on other persons (i.e., being self-reliant), or without substantial assistance from outside resources (govt., etc.) in order to minimize the effects of that change on their current lifestyle.

To be a real prepper, one needs to look at the situation from a different perspective; one needs to be a believer in Jesus Christ and put his or her faith in Christ. All the bunkers, caves, gold, guns, and groceries in the world cannot save anyone from the coming catastrophic events. We need to develop a personal relationship with Jesus Christ, and we should study the prophecies of our Bibles and how they relate to current national and global trends. We should take comfort in the knowledge that God has made available to us about end times. When we prepare ourselves spiritually, it will be a time of confidence, hope, and joy, which will ultimately lead to God's promised eternal kingdom.

The world's principal religions and spiritual traditions may be classified into a small number of major groups. Near the beginning of the nineteenth century, between 1780 and 1810, the language dramatically changed; instead of "religion," authors began using the plural "religions" to refer to both Christianity and other forms of spirituality.

The CIA World Factbook of 2010 reports that adherents of the world's religions consist of 33.4 percent Christians, 22.7 percent Muslims, 13.8 percent Hindus, 6.8 percent Buddhists, and 11.6 percent "other," with 11.7 percent of respondents identifying as nonreligious. This does not represent a comprehensive list of all religions, only the major ones, but they account for more than 98 percent of the world's population. The nonreligious group includes atheists, agnostics, secular humanists, and people that say they have no religious preference. About half of this group does believe in some sort of deity, but they prefer to be classified as nonreligious.

Fundamentally, the world's population can be divided into two groupings: there are born-again Christians, and there are those who are not born-again Christians. Born-again Christians can be categorized as the Christians that accept the existence of a supernatural God (the God of the Bible), have accepted Jesus Christ as their personal savior, and are anticipating eternity in heaven, as promised in the Bible, after they die. The people who are not born-again Christians include adherents of Christian denominations and other religions who have not specifically accepted Jesus Christ as their personal savior. They also include agnostics and atheists. Agnostics are the ones who just live their lives, have no consideration for the afterlife, and do not give a hoot if there is a heaven or not. Then there are atheists.

What is an atheist? There are many different levels of atheism. Implicit atheists include people such as young children and some agnostics, who have never been introduced or made aware of the existence of a God but have not explicitly rejected such a belief. In 1979, American writer George H Smith coined the term "implicit atheist" to refer to "the absence of theistic belief without a conscious rejection of it." An explicit atheist is someone who believes that "At least one God exists" is a false statement. An explicit atheist believes that man is just another living biological being, like all other living organisms on earth. He does not believe that a human being actually has a soul, and believes that when he dies everything comes to a finale, over and done with, and that there is nothing further. Then there is the materialistic atheist, like one of my colleagues, who simply states, "I have no need for a God. I have plenty of money; I can buy whatever I want to make me happy. I choose gold before God." Sadly, he does not realize that eternal happiness in paradise cannot possibly be bought with gold, money, or material wealth. Often, explicit atheist defend their positions with statements like "There is no God, because I say so." Some simply state, "I don't bother with this nonsense; I am a lot smarter than that. Religion is only for losers"! However, it is also a well-known fact that atheists are quick to convert and become believers in times of severe crisis, and that they often make last-minute deathbed conversions to Christianity. And it is certain that atheism has no hope beyond a cold hole in the ground. According to one estimate, atheists make up only about 2.3 percent of the world population. Rates of self-reported atheism are among the highest in Western nations, varying from 4 percent in the United States to 32 percent in France. However, if people claim to be nonreligious that does not necessarily mean that they are explicit atheists. They could very well be born-again Christians who do not attend regular church services or belong to a religious organization.

I have also heard many people say, "Well, I have been a good person all my life, I have lived a good life, never cheated anyone, always been considerate of others, and if that is not good enough, then so be it." However, it is fine and wonderful to be such a noble person, but it is just not good enough; it takes just a little more than that. The Holy Bible, or the Word of God, makes it very clear what one must do in order to be rewarded the promised paradise. We must realize that we are not the ones that set the criteria for going to heaven and that we are not the ones that dictate the rules. Only Almighty God Himself sets the criteria for going to heaven, and we are not the ones that can take authority away from Him and make our own rules. Ultimately, according to the Bible, believers, or people who have accepted the fact that God and heaven do exist, will be eligible for eternal happiness and joy in heaven, and the ones that refuse to accept God will be reprimanded with everlasting pain and regret. "I am the Lord, and there is no other; apart from me there is no God. I will strengthen you, though you have not acknowledged me" (Isaiah 45:5). "For what doth it profit a man, to gain the whole world, and forfeit his life? For what should a man give in exchange for his life? For whosoever shall be ashamed of me and of my words in this adulterous and sinful generation, the Son of man also shall be ashamed of him, when he cometh in the glory of his Father with the holy angels" (Mark 8:36–38).

What is the purpose and meaning of our life and existence here on Earth? Why has God created us? The answer to this question has been the subject of debate for many centuries. The sole purpose of us being here on Earth is to prepare our soul for the final eternal destination—either the kingdom of heaven or, alternatively, the horrors of hell. According to the Bible, these are the two only options available to us and we must decide to spend our spiritual life after death in either one or the other. A believer makes this crucial decision while he is still alive and he knows he will spend eternity with God in paradise; a nonbeliever must wait until the moment of death before he finds the truth. The Bible is not very specific as to what heaven really is like; it says only that it is a place where there is no pain, no tears, and no hunger or thirst, and that its streets are paved with gold. Heaven is a place of eternal happiness and joy. Heaven is the place where we will reunite with our friends, relatives, and loved ones that have gone before us.

Most people believe that hell is a place reserved only for the most hardened criminals and most evil elements of our society. However, the Bible clearly states that only those who purposely reject God's merciful gift of salvation will spend eternity in this place, regardless of all the good deeds and good works they may have done. It is not just good deeds that will bring one to heaven; neither is it bad deeds that will bring one to hell. The Bible continually warns of this place called hell. Hell is a place for punishment and suffering during the afterlife; it is reserved for people who choose to reject Christianity or refuse to accept Jesus Christ as their Lord and Savior. There are over 162 warnings regarding hell in the New Testament alone. Moreover, over seventy of these references were uttered by Jesus Himself! However, the word "hell" appears only thirteen times, in thirteen verses, in the New American Standard Bible. The Bible does not elaborate on the conditions in hell other than referring to it as Hades, the everlasting lake of fire, or "a place of everlasting torment"—"torment" meaning "torture"; neither one seems very appealing. The atheist, however, not believing in God or even considering the possibility of having a soul, consequently will never have even the slightest chance ever to see heaven; he will always be on the losing end. Unlike the believer, there are no chances for him ever to be a winner. I believe the worst punishment the nonbeliever will endure is the eternal (everlasting) regret that during his short life here on Earth he has consistently refused to accept the gift of the salvation of Jesus Christ. At the moment of his death, he will come to realize that he is too late; that he has blown his chance and now has to suffer the harsh consequences of his remorseful decision for eternity. There will never be an end to his suffering—never.

The believer, on the other hand, has already predetermined his final destination; he knows he will enter the kingdom of heaven. The other place is no option for him; he knows that he can never be a loser!

> ... even the righteousness of God through faith in Jesus Christ unto all them that believe; for there is no distinction; for all have sinned, and fall short of the glory of God; being justified freely by his grace through the redemption that is in Christ Jesus: whom God set forth *to be* a propitiation, through faith, in his blood, to show his righteousness because of the passing over of the sins done aforetime,

in the forbearance of God; for the showing, *I say*, of his righteousness at this present season: that he might himself be just, and the justifier of him that hath faith in Jesus. (Romans 3:22–26)

What if there were no God? If there were no God, there would be absolutely nothing at all; we would not be here, our precious earth would not be here, our universe would not be here, and no, there would be no heaven or hell either. However, we are here, our earth is here, and our universe is here; this means that God is also here! If we were here without the existence of a God, every one of us here on Earth would live without a purpose. Our final destination would be the grave, where our bodies would decompose and return to the same earth we came from in the first place. There would be nothing left, not even a bit of sadness or a pleasant memory. At the end of our lives there would be no reward, no reprimand, absolutely nothing; we would have lead our lives without any meaning, or purpose, or logical reason for us to be here in the first place. "The fool hath said in his heart, there is no God. Corrupt are they, and have done abominable iniquity; There is none that doeth good" (Psalm 53:1).

"Faith" is another word for believing in a God or a superior supernatural being that cannot be scientifically explained. However, it does require substantially more faith not to believe in the existence of a God than it does to believe. Anybody and his dog can easily prove that there indeed exists a God, but there is no possible way of proving that God does not exist. Just because God cannot be physically seen or scientifically explained does not mean that He does not exist. To prove God's existence, all we need to do is take a good look around us and observe the environment that surrounds us and the wonderful miraculous things that He has created for us, some of which even modern science still cannot explain to this very day!

What happens the minute after we die? The precise moment we die is also the precise moment our souls separate from our bodies. The body slowly starts the process of decomposition and returns to the dust of the earth, just as the Bible says. The destination of our soul depends on the choice we as an individual made while we were still united as one unit, body and soul.

There are actually thousands of well-documented cases in which people have died and gone beyond—looked behind the curtain, so to speak—and then were given the opportunity to return to Earth. The stories they tell are almost identical in just about every case. They usually can describe every minor detail after the moment they die; first they hover over their body, then they enter some sort of tunnel through which they move at incredible speed. Their entire life flashes before them in just a matter of what seems like seconds as they remember every minute detail of their entire life, and at the end of the tunnel is a tremendously bright light, described by some as being "a thousand times brighter than the sun." They all seem to conclude that this bright light is nothing other than a divine presence, because they describe it as an immensely intense feeling of love, peace, and joy beyond human comprehension. Together with this intensely positive feeling, these people also report a tremendous increase in awareness and consciousness, as if they were experiencing a taste of heaven. For them, there is always some point of decision, some point of no return; either they must go through a door or a gate, or cross a bridge to get to the other side, but they always seem to realize that once they cross the threshold, they will never be able to return to Earth. Interestingly, almost all people who have had such experiences become born-again Christians.

Joyce Hawkes had an accident that forever changed her life and her view of science. She suffered a concussion from a falling window. "I did not bump the mantel or touch the fireplace, yet in a flash a heavy, leaded-glass art piece framed in thick oak toppled off the mantel onto my head. I crumpled to the carpet, crushing pain shot through my head, and I was out. From there on nothing was ordinary. My awareness no longer resided in my body. I had no sense of a body, or of my house, or of living across the street from Greenlake in Seattle, or even of my own name. I was zipping along a long, dark tunnel, drawn to a beautiful and welcoming light far ahead. At the end of this tunnel, just before the entrance to the lighted place, my deceased mother and grandmother stood. They were radiant with good health, glowing with love, and welcomed me wordlessly. I was overwhelmed to see them. I had missed them intensely, but had no belief in an afterlife, so seeing them astonished me. It seemed like we were together for an eternity, and yet I moved on without remorse or sadness into the place where the light was stronger still." "I think that part of me, my spirit and my soul, left my body and went to another reality." She was surprised at the experience. "It just was not part of the paradigm in which I lived as a scientist. I think what I learned was that there truly is no death, that there is a change in state from a physical form to a spirit form, and that there's nothing to fear about that passage." Joyce Hawkes, Ph.D., is a biophysicist and cell biologist. She completed her doctorate in biophysics at Pennsylvania State University and was a postdoctoral fellow with the National Institutes of Health before settling in Seattle to work in research for the National Marine Fisheries Service, a part of the National Oceanic and Atmospheric Administration. While there, she was honored with a National Achievement Award for her work.

The Bible tells us that God is Love, and perhaps this is what Jesus was referring to when He spoke the words "I am the Light." (John 8:12)

For the believer, or born-again Christian, the answer is very simple. He already knows what is going to happen; he is well aware of the consequences of his physical death and his spiritual destination. He knows his demise is not the end; it is only the beginning of eternity in paradise. The Jewish leader Nicodemus asked Jesus, "What does it take to enter the Kingdom of God?" Jesus answered and said to him, "Verily, verily, I say unto thee, except one be born anew, he cannot see the kingdom of God" (John 3:3).

What does it mean to be born again? Born-again Christians are people who have repented of their sins and turned to Christ for salvation. They are people who believe in God and have accepted the fact that Jesus Christ was indeed the Son of this God and came to Earth to redeem all the sins of the world, including theirs. The Bible clearly states that everybody is naturally a born sinner. What this means is that we are not sinners because we commit sin; rather, we commit sin because we are born sinners! By nature, we are not members of God's family and have no right to inherit eternal life. Because of our sins, we have rebelled against God, and God separated us from Him. There is no possible way we can earn forgiveness for our sins; nor can we earn eternal life through any means other than accepting the salvation of our Lord Jesus Christ; there is absolutely nothing else we can do that can get us off the hook and unlock the gates of heaven. All good intentions and good deeds are just not sufficient. Thankfully, God, though perfect righteousness, is also wonderfully gracious and merciful. Because of His love for us, He sent His only Son, Jesus, to take the punishment for our sins and to die as a criminal on the cross at Calvary. Jesus Christ is God, and He paid the ultimate sacrifice so that our sins may be forgiven and we may have everlasting life. We must sincerely accept in our hearts the fact that Jesus is God, and we must put our faith in Him and ask Him to forgive our sins. Once we have truly committed our lives to Christ, we know we are born-again Christians and eternal happiness is awaiting us, without the shadow of a doubt. "… having been begotten again, not of corruptible seed, but of incorruptible, through the word of God, which liveth and abideth" (1 Peter 1:23).

For the atheist, however, things are quite different. The biggest surprise he will get is, at the moment of his death, the realization that he is not really dead after all, but that he is very much still there and looking down on his body, which has already started the slow process of decomposition and being returned to the dust of the earth. He will slowly start to realize that as a human being he actually did consist of a body and a soul, and that his body at this point is nothing more than a lifeless pile of garbage, while he, his spirit, is still present and looking down at it. Of course, after realty sinks in, the big "What if" question slowly starts to creep in. *What if all these people talking about God were actually right and there really is a God? What if heaven really does exist? What if the thousands of churches I have been driving by during my lifetime actually did have a purpose and a meaning? Perhaps I should have taken a little time to investigate. What if all those TV preachers were actually right in proclaiming that there is a God and that their message had a little more meaning than just begging for money?* Then, slowly but surely, the next question will come about. *Now what?*

Realizing that he is still here and not everything is over yet, he will come to realize that he does have a soul after all, and suddenly the destination of his soul will become his greatest concern. His next reaction, of course, will be, "Oh my God, now what"? He will rather quickly realize that perhaps he should have stopped at one of those churches and asked some questions, or paid more attention to those TV preachers and listened to what they had to say. Suddenly it will sink in that now is the time to change his mind, while he thinks he still can. However, he will sadly arrive at this decision just a little too late, and although he is looking down on his body, he will realize that his brain is no longer active and no longer functioning; his knees will no longer bend, and his mouth can no longer speak the words. He will be too late indeed. He will realize that he has blown his chance and that he should have made this decision just a little sooner.

Oh my God, now what? He will suddenly become aware that because of his denial and defiance of God he will never be allowed to enter the kingdom of heaven, and consequently he will slowly start to wonder about the other place. He will realize that he will spend his second life—his eternal life—in "not such a nice place." At this point, he will probably just wish he could kill himself, be gone, and get it over with. But ultimately, he cannot kill the soul, and his biological body is already dead. Like it or not, he is stuck where he is and he has to continue with the consequences of his decisions for eternity. "And be not afraid of them that kill the body, but are not able to kill the soul: but rather fear him who is able to destroy both soul and body in hell" (Matthew 10:28).

For the atheist who confessed that he needed no God, who thought he had enough money and wealth to buy whatever he wanted, there is also a great surprise in store for him. He will realize that his money or wealth is no longer available to him and cannot possibly buy his way into heaven; nor will it buy his way out of hell. He can no longer live in his fancy villa, drive one of his fancy automobiles, or access one of his bank accounts. He will suddenly remember what he learned at Sunday school ages ago—that the salvation of God is a free gift; it cannot be bought with either money or good deeds. He will slowly realize he has arrived at a place where all his money and wealth, always so precious to him, have absolutely no meaning and absolute zero value. However, he will sadly arrive at this conclusion too late, slowly remembering that Jesus Christ already bought his salvation for him some two thousand years ago, by sacrificing His life and dying on the cross for all people of the world, and that all he had to do was just ask for salvation in order to receive it. He will realize that it is what is in your heart that will bring you to heaven, not what is in your wallet! "For it is easier for a camel to enter in through a needle's eye, than for a rich man to enter into the kingdom of God" (Luke 18:25).

The most significant point of the entire Bible is the crucifixion of our Lord Jesus at Calvary, simultaneously with two hardened criminals—the scum of society, so to speak—one on His left, and the other on His right. The one on the left turned to our Lord and said, "Hey you, if you really are who you say you are, come on, do something; get us out of this mess!" On the other hand, the one on the right showed some remorse, realized who his neighbor on his left was, and slowly lifted his head and said, "My Lord and my God, please forgive me." Jesus lifted his head, looked at him and answered, "Verily I say unto thee, To-day shalt thou be with me in Paradise" (Luke 23:43).

The difference is that the one on the left asked Jesus to perform a miracle and save his body, while the one on the right asked Jesus to save his soul. The victim on the left obviously died as a criminal and an atheist, never to see the kingdom of heaven. The results for the one on the right, however, turned out to be quite different; he, in the very last moments of his miserable life, accepted Jesus as his savior and asked Him to forgive his sins, and thereby he became a born-again Christian. The pathway to heaven was cleared for him, and all obstacles were removed from his way. Someday soon I am certain to meet this person; the nonbeliever, on the other hand, is likely to meet the other guy, at some other location!

The exact same option is available for each and every person that lives on the face of Earth today, even you! The only criteria for becoming a born-again Christian are sincerely accepting the grace of our Lord Jesus in your heart and asking Him to forgive your sins. At the exact moment you do this, you become a saint, or a born-again Christian. At this precise moment, your sins and all your transgressions are instantly forgiven and shall never be remembered. All your skeletons are removed from your closet, so to speak. At this exact moment, the gates of the kingdom of heaven are opened for you and you are welcomed inside!

Regardless of what kind of criminal, crook, or louse you might have been during your life on this earth, from the moment you accept Jesus Christ into your life and ask Him to forgive your sins, you continue with a perfectly clean slate. This option is available for each and every one of us, including the hard-core atheist and all other nonbelievers; however, it is of uttermost importance we make this crucial decision while we are still alive here on this earth, before the moment our soul separates from our body! After that, we are simply too late; we can no longer change our minds. And is it really relevant at which point in our life we make the crucial decision to accept Jesus as our Savior? What if it is the very final moment of our life? Remember the criminal at the right hand of Jesus on Calvary; he was saved during the very last moment of his life and eventually was rewarded the kingdom of heaven.

Society, or the state for that matter, may not be so lenient. Society and the state tend to be a little more hard-nosed, and they will never forget. Once you have committed a crime, the record of your crime will follow you wherever you go for the rest of your life and beyond, even though you have served the time for your crime, apologized, and shown remorse. The state keeps a permanent record of your repulsive past, passing it on to future generations; but that really is of no consequence, since the state does not have authority over the gates of heaven—only God does!

God promises that once your sins are forgiven He will never remember, meaning that in paradise there will be no recollection of any of your troubled and sinful past. Memory is part of a person's brain; when the body dies, the brain also dies, completely eradicating the recollection of any negative events that occurred during a person's life.

> And their sins and their iniquities will I remember no more.
> (Hebrews 10:17)

Be it known unto you therefore, brethren that through this man are proclaimed unto you remission of sins: and by him every one that believeth is justified from all things, from which ye could not be justified by the Law of Moses. (Acts 13:38–39)

… for all have sinned, and fall short of the glory of God. (Romans 3:23)

For the wages of sin is death; but the free gift of God is eternal life in Christ Jesus our Lord. (Romans 6:23)

Because Christ also suffered for sins once, the righteous for the unrighteous, that he might bring us to God; being put to death in the flesh, but made alive in the spirit. (1 Peter 3:18)

And they said, Believe on the Lord Jesus, and thou shalt be saved, thou and thy house. (Acts 16:31)

And in none other is there salvation: for neither is there any other name under heaven that is given among men, wherein we must be saved. (Acts 4:12)

Jesus saith unto him, I am the way, and the truth, and the life: no one cometh unto the Father, but by me. (John 14:6)

These things have I written unto you, that ye may know that ye have eternal life, *even* unto you that believe on the name of the Son of God. (1 John 5:13)

Marvel not that I said unto thee, ye must be born anew. (John 3:7)

Blessed *be* the God and Father of our Lord Jesus Christ, who according to his great mercy begat us again unto a living hope by the resurrection of Jesus Christ from the dead. (1 Peter 1:3)

…having been begotten again, not of corruptible seed, but of incorruptible, through the word of God, which liveth and abideth. (1 Peter 1:23)

So, my friend, be wise. Bend your knee and gracefully accept the salvation of the Lord Jesus in your heart. Ask him to forgive your sins and trespasses, and I will see you in paradise, perhaps much sooner than we both think! What you are about to gain is eternal happiness and joy, while there is absolutely nothing to lose! "I say unto you, that even so there shall be joy in heaven over one sinner that repenteth, *more* than over ninety and nine righteous persons, who need no repentance" (Luke 15:7).

Index

Transparency International, 184
Transubstantiation, 192–196
Tribulation, 148, 149, 151, 152, 153,
    155, 171, 181
Twin Towers. *See* 9/11 terrorist attacks;
    World Trade Center
Tyndale, William, 193

U

UFOs, 156–157
Ultraviolet radiation, 84
Unam sanctam, 188–189
Universal gravitation, 122–123, 125, 131
Universe
    age of, 1
    expanding, 9–21, 106
    heliocentric cosmology and, 2–7
    origins of, 1–2, 12, 23, 27–36, 42
    size of, 24
    See also Big Bang Theory; Steady
        state theory
The Unrandom Universe (Brouwer), 120
Uranus, 114–115
Uriah, 183–184

V

Van der Woude, Jurrie, 117
Van Flandern, Thomas, 19–21
Van Impe, Jack, 139, 143, 177, 178
Vatican, 200
    See also Catholic Church
Venus, 111
VeriChip, 162
Villard, Ray, 117
Virchow, Rudolph, 67
Visible light spectrum, 82
von Braun, Wernher, 38
Vos Savant, Marilyn, 29
The Voyage of the Beagle (Darwin), 44
Voyager 1, 117
Voyager 2, 115, 116

W

Walvoord, John F., 139
Warren Commission, 200–202
Watchmaker analogy, 51–52
Water, 120
Whirlpool Galaxy, 100–104
Whisenant, Edgar C., 142
Wilder-Smith, A.E., 54
Wilson, Robert, 15, 16, 86
Winterberg, Friedward, 128–129
Woodward, Author Smith, 68
Woodward, Bob, 210
Word of God. *See* Bible
World government, 163–164
World Trade Center, 204, 212–215
    See also 9/11 terrorist attacks
World Zionist Congress, 172–173
Wright, Thomas, 97–98
Wyatt, Ron, 183
Wycliffe, John, 193

X

X-rays, 84

Y

Y2K bug, 140–141
Yom Kippur War, 173
Young Earth Creationists, 22, 48, 50,
    54

Z

Zionist movement, 172–173
Zuckerman, Solly, 69
Zwingli, Huldrych, 193

Gary Haitel was born in Emmen, Netherlands, and immigrated in 1968 to Canada, where he pursued a career in cabinetmaking and interior architecture. Gary grew up in a very strict Roman Catholic family and later became a nondenominational Christian. Mr. Haitel has always expressed an interest in science and astronomy and has devoted most of his adult life pursuing the reality of the natural world and its relationship to the Holy Scriptures. He has attended many courses and studies and has spoken on the subject on several occasions. Mr. Haitel now lives in Calgary, Alberta, Canada.